City and Environment

City and Environment

CHRISTOPHER G. BOONE
AND ALI MODARRES

TEMPLE UNIVERSITY PRESS
Philadelphia

Temple University Press
1601 North Broad Street
Philadelphia PA 19122
www.temple.edu/tempress

⊗ The paper used in this publication meets the requirements of the American
National Standard for Information Sciences—Permanence of Paper for
Printed Library Materials, ANSI Z39.48-1992

Library of Congress Cataloging-in-Publication Data
Boone, Christopher G., 1964-
    City and environment / Christopher G. Boone and Ali Modarres.
      p. cm.
    Includes bibliographical references and index.
    ISBN 1-59213-283-9 (cloth : alk. paper) — ISBN 1-59213-284-7 (pbk : alk.
    paper)
      1. Urban ecology. 2. Sociology, Urban. I. Modarres, Ali. II. Title.
HT241.B66 2006
307.76—dc22

2005055920

2 4 6 8 9 7 5 3 1

To our families

Contents

List of Tables and Figures

Preface

IN A RECENT ISSUE of *The New Yorker*, Manhattan was described as the greenest city in the United States. Dense population, high land values, and historical legacies of development have created a city that is difficult for driving but great for public transit, nearly impossible for detached housing but well suited to smaller apartments, terrible for parking but ideal for walking, rotten for big box stores but perfect for corner shops in residential neighborhoods. The concentration of housing with no backyards created demand for public parks, most notably Central Park. Infrastructure, while difficult to construct, is extremely efficient compared to sprawling suburbs. All of these characteristics mean the amount of energy and materials required to support a person living in New York is considerably smaller than that required to sustain someone living in a typical suburb. New York, in this respect, could serve as a model of sustainable development (Owen 2004).

That a city, and even a fantastically large metropolitan area, could be a model of sustainable development may provoke gasps of horror from some people. American environmentalism has a long history of wilderness and wildlife protection and of seeing cities as the ultimate expression of human domination over nature. "Nature" is what city dwellers find beyond the city limits sign. The problem with this way of thinking is that it neglects the natural or nonhuman organisms that exist in cities, even in the densely built-up cores of urban areas. Trees do grow in Brooklyn. In addition, this arbitrary distinction between city and nature separates human beings from other organisms and from the environment in which they live. The city/countryside, urban/rural, human/natural dichotomies are beginning to become blurred in environmental science, but they still resonate in the minds of many. Ask anyone to draw sharp boundaries between the urban and the rural or between the human and the natural world, and the difficulty with these concepts becomes plain. The ecological structure and function of midtown Manhattan are certainly different from the patterns and processes observed in boreal forests, but there are gradations between these two types of places. It may be better to think of urbanness, or the attributes that make a place urban, rather than defining strict boundaries between cities and countryside or wilderness.

Cities have ecologies. They are special types of ecologies dominated by the actions and products of human beings. Ecology, the science of relationships between organisms and their environments, is most often applied to the nonhuman world, where the behavior of organisms such as trees or worms is predictable and systematic. But the basic principles of ecology, coupled with observations and theory from the social sciences, can be applied to understanding human beings and their relationship with the environment. This work is only now maturing, and it is still a small part of the larger field of environmental studies. For instance, only two sites out of twenty-six being studied as part of the Long Term Ecological Research (LTER) network in the United States are in metropolitan areas—Baltimore, Maryland, and Phoenix, Arizona.

Social scientists too are working to understand how ecosystems and human beings affect each other. This "social-ecological systems approach" tries to integrate the biological and geophysical processes studied by ecologists with what we know about human behavior to better understand human and ecological patterns and how these processes function. Despite formidable obstacles, the social-ecological approach is laudable, as it pulls ecologists away from the typical characterization of humans solely as "disturbers" and pulls social scientists away from the notion that human beings (and their cities) are driven solely by culture and divorced from ecological processes (Redman et al. 2004).

We have written this book not only to illuminate the social-ecological systems of cities as an academic exercise, worthy as that may be, but ultimately to help in the search for pathways that can turn cities into ecologically sustainable places. Given the human penchant to consume vast amounts of resources and produce large quantities of waste, we need to look for real, practical solutions for meeting human needs while conserving or enhancing natural resources, biodiversity, and overall ecosystem health. Readers will find discussion of both the current environmental state of cities as well as ideas such as "sustainable development," an approach that academics, decision makers, nonprofits, and others have adopted or championed because it offers a win-win solution, promising economic prosperity and environmental improvement. Sophisticated versions of this model embrace the idea that human well-being (jobs, food, sanitation, health, energy, wealth) is ultimately dependent on environment but that the relationship is reciprocal. While we are dependent on the earth for our needs, we must at the same time remake that environment in such a way that it is able to support human well-being (Giddings et al. 2002).

But sustainable development is hampered by uneven development and inequity. Representatives at the World Summit on Sustainable Development, held in Johannesburg in 2002, deemed the persistence of poverty a major obstacle to sustainable development. Poverty usually means fewer choices, short-term thinking, and poor health, all of which undermine the basic prescriptions of sustainable development. In cities, the benefits and costs of urban life are typically uneven. Besides gaping disparities in income, one issue that has attracted considerable attention is the inequitable or uneven distribution of waste dumps and toxic sites and other environmental disamenities. The pursuit of a fair and equitable distribution of these disamenities, so that no single group bears a disproportionate burden, has often been called the search for environmental justice. Typically the focus has been on minority, low-income, and less educated groups, which tend to wield less power in decisions that affect the location of unwanted land uses. Traditionally, environmental justice has focused on disamenities, but it can also address the lack of amenities, such as green spaces or street trees, that have a direct impact on social and individual well-being. It is an idea that has found currency in activism as well as in scholarship. Community activists have been able to use environmental justice arguments to improve environmental conditions in neighborhoods.

Another complication in the effort to create sustainable and livable cities, developed in the pages that follow, are the differences between the so-called brown and green agendas. The green agenda tends to favor environmental policy and action that protects or enhances nature over the needs and concerns of human beings. The brown agenda focuses on the environmental problems faced by human beings, such as pollution, poor sanitation, and exposure to toxins. While proponents of the brown and green agendas recognize the linkages between ecosystem and human health, they differ over which issues should receive priority. Regarding solid waste, the brown agenda focuses on the need for regular collection of household waste, especially in poor neighborhoods in developing countries, where such collection is a rarity. Improper disposal of waste in rivers or streets can attract vermin, spread disease, and degrade human health. From the green agenda viewpoint, the focus should be on recycling, waste reduction, and reuse of materials rather than on investing in waste removal. Scholars have noted that the brown and green agendas can be reconciled, and in a way that is efficient and effective. Waste management systems that provide both service to all and the means to recycle and reuse materials can improve community health and provide jobs while reducing resource consumption, protecting waterways, and slowing the waste stream (McGranahan and Satterthwaite 2002). The waste-for-transportation-tokens program in Curitiba, Brazil (Chapter 5) shows how smart planning can lead to win-win solutions, satisfying proponents of both brown and green agendas. Above all, creative thinking is a key ingredient in making cities sustainable.

This book critically examines urban environmental issues for cities, past and present, rich and poor. We begin by reviewing urban morphology and the construction of ideal cities in human history. In Chapter 1 we examine what cities have meant to us and what we have expected from them. Focusing on space and meaning brings us closer to situating urban design and governance within the larger framework of current debates on sustainability, which we interpret in Chapter 2, on the creation of an urban ideal. This chapter sets the stage for a series of discussions on cities and their environments. We tackle the issue of population and attempt to situate this discussion, once again, within sustainability and environmental equity issues. Chapter 2 offers alternative ways of thinking about population dynamics and policies that attempt to control population growth, through specific examples from various regions of the world.

Any discussion of population must take into account the question of sustainable food supplies. In Chapter 3 we turn to agriculture, which binds human and ecological systems and makes human life and city life possible. This chapter draws attention to the irony that farms feed cities while urban growth threatens agricultural hinterlands. It also looks at urban sprawl as a clear physical manifestation of consumption as well as the far-flung ecological footprints that result from consumption and waste production in cities. Finally, it examines the increasing practice of farming in cities, or urban agriculture.

In Chapter 4 we illustrate how urban infrastructure has been used to overcome the environmental limitations of crowded cities and how this has exacerbated other environmental and health issues. Green alternatives to hard or gray infrastructure, such as artificial wetlands for water purification, are also examined. Chapter 5 begins with an assessment of major health issues in cities and current approaches to improving health. It examines the role of policy as well as of design in making cities healthier, both in developing and developed countries.

The final chapter tackles the issue of green space in cities. Given the importance of past decisions on present-day green space, this chapter focuses on the historical development of parks and open-space philosophies. It ends with a critical look at recent planning ideas, including smart growth and new urbanism, and concludes with some suggestions for creating sustainable cities.

Acknowledgments

THE PRODUCTION OF any book requires the attention and support of many individuals. We especially thank our thoughtful reviewers, whose comments and insights helped us improve the content of the book, Peter Wissoker, who helped us get through the process, and Alex Holzman for seeing us through the completion of the final manuscript.

We are grateful to Marcia Nation, who, drawing on her thesis on British greenbelt planning, contributed some text in Chapter 6. Marcia also offered a careful and thoughtful critique of many chapters, and we thank her for that. We also offer our thanks to Andrea Modarres, who read some of the chapters and made valuable comments.

No academic book is created in an intellectual vacuum. Many of our ideas were influenced by our mentors, colleagues, and students, whose inquisitive minds inspired us. We also wish to thank our academic institutions for their support in giving us the time to develop the manuscript and promoting our academic aspirations.

1 Urban Morphology and the Shaping of an Urban Ideal

CITIES ARE the greatest of human inventions. They embody our histories and manifest our technological innovations, cultural and social interactions, economic structures, political systems, and our respect for (or fear of) deities. Cities contain our imagined communities, our socially constructed identities, and the spaces that shape our daily activities. We equate cities with progress, and in many cases cities elevate their citizens to higher social status than that afforded to their rural counterparts. As representations of who we are (and who we were), cities have been the objects of our desire, our love, and our hate. With mixed emotions we have come to imagine them as sites of comfort and safety as well as of poverty and misery, filled with vice and immorality and godlessness. Cities are the nexus of production and consumption, service provision and neglect. These dichotomies have been with us since ancient times. Judeo-Christian texts assailed the magnificent cities of Mesopotamia as places unfit for "true believers." According to these texts, God accepted the farmer's gift, burned Babylon, and ran Lot and his family out of Sodom and Gomorrah. But if God frowned upon the corruption of cities, their denizens loved them, nourished them, and handed them down to us—their urban descendents—as gifts. Not unlike them, we seek in cities what cannot be attained in the country—anonymity, economic mobility, social ascendance, money-ordered civil society, and the ability to produce and consume at the highest level possible. It is in cities that civilizations have been born and it is in cities that the concept of the citizen was born.

We seek evidence of culture, social organization, economy, and politics in the physical structure of cities. Documenting history primarily from an aesthetic perspective, cities are our most persistent spatial narratives. They predate nation-states and in all likelihood will outlast them. Despite the obsession of architectural historians and historical texts whose purpose is to further an exaggerated sense of nationalism, cities are as much about the everyday life of their residents as they are about monuments. Though the life, contribution, and spatialized existence of elites may be the mainstay of most urban histories, we agree with James Vance (1990) that the least studied spaces of the city belong to vernacular architecture. For cities to be understood in their totality, urban form cannot be limited to social, political, and economic processes but must include the physical structure that defines its spatial connectivity and hierarchy. Why urban forms persist despite revolutions but change in response to specific technological innovations is at the heart of how cities survive the upheavals that occasionally rattle the sociopolitical organizations that they house. Cities are not simple pawns in an economic and political game of chess. Spatial forms and their morphogenesis cannot be understood through political economy alone; our understanding also must be informed by everyday existence, alienation, resistance, and resilience. After all, cities of people clearly have outlived the cities of gods, god-kings, and demigods. Though certain elements of the urban form may have been created through the direct intervention of political and economic elites, it is the city of people that defines the urban experience.

In this chapter we review concepts in urban morphology from historical, social, political, economic, and planning perspectives. First, we trace urban morphological patterns in various regions of the world. Our intention is to help readers derive a regional, or at least a temporal, understanding of distinct urban typologies. We discuss briefly the various phases and types of urban forms, including early cities, Greek, Roman, Muslim, medieval, Renaissance, industrial, and postindustrial cities. We then discuss the role of modernity in altering the shape of cities. And, finally, we review some of the current debates on urban design and alternative ways of conceptualizing the physical structure of cities.

INVENTING CITIES

Most urban histories begin with a discussion of the Agricultural Revolution and its impact on the birth of cities. This history is then tied geographically to Mesopotamia. The eminent cultural geographer Carl Sauer (1952), however, argued that the practice of agriculture might have occurred first among sedentary populations in Southeast Asia—not in Mesopotamia—fourteen to thirty-five thousand years ago. This theory, which has been supported by archeological investigations, is based on the deduction that vegetative reproduction (agriculture that does not rely on seeds but employs roots and plant branches to grow and reproduce plants) may have preceded the more sophisticated agricultural practices found in Mesopotamia. Conflict over resources, management of irrigation systems, development of labor specialization, and the need for administrative functions that were beyond the capacity of individual farmers were also responsible for the emergence of what we recognize today as urbanization. It is no surprise that the godlike heads of the early city-states rarely took the image of an ascended farmer but rather assumed the figurative power of a hunter. What is important for us to realize is that the history of urbanization is not a simple transformation of hamlets to cities, based entirely on the addition of more people. The birth of cities is connected with innovations in agriculture and also, more significantly, in governance and the management of highly concentrated populations.

As many others have done, we identify five urban hearths in the world, from which urbanization diffused. These are in Mesopotamia (in the flood plains of the Euphrates and the Tigris), Pakistan (along the Indus River valley), China (in the valleys and alluvial fans of the Huang River), Egypt (along the narrow flood zone of the Nile), and Mesoamerica (including, according to some scholars, Andean cultural centers). Proximity to fertile soils, delivered annually by river floods, made these places ideal for the birth of cities. Morphologically, then, it should come as no surprise that, along with the dependable agricultural production in close proximity to these sites, the earliest cities would dedicate special areas to the storage and management of food. A city devoted to agriculture would also build and maintain irrigation systems, make rules for the distribution of goods and water rights, erect walls to defend the stored food (and those who protected the food—the ruling class), and invent deities concerned with agricultural productivity. The code of Hammurabi is a testament to these innovations and to the system that protected and maintained the cities. The emerging urban cultures, with their dependence on agricultural productivity, would seek in king and gods the protection of their common well-being.

Before the God of Abraham attracted a large following, fertility gods were venerated in city and country alike. In cities they were monumentalized in impressive temples and walled buildings. Echoing these so-called pagan gods' role as the common deities of farmers and city dwellers, the Semitic religions pointed to Cain, the son of Adam and the original farmer (who killed his brother Abel, the herder) as the founder of the city of Enoch, after the name of his son. Interestingly, the Semitic religions adopted a lunar calendar that closely corresponded to the life-style of farmers.

Agricultural order was invented in cities and imposed by their rulers on farmers. Villages did not turn into cities; they gave birth to physical and social systems that came to rule, regulate, and subjugate them. The city-states of Mesopotamia were made possible by the order wrought on agriculture by urban governments and were ruled by godlike kings whose interest in sustaining their power enticed them to build defensive structures and protect both cities and the agricultural networks that fed them. But how did city-states appear, and why? What were the first cities like? And what do we know about their morphology?

Maisels (1990) offered a model of urban development (i.e., city-states) that may start us on a path to answering these questions. Building on the hydraulic civilization model, which argues that a sophisticated irrigation-based agriculture was a prerequisite for the formation of city-states, Maisels suggested that conflict and redistribution may be used as factors for understanding how cities were spun into existence. He used Wright's (1978) extension of Carneiro's (1970) external conflict theory to illustrate the process by which states could have originated. In this model, population growth and circumscribed agricultural resources contribute to warfare, which leads to the formation of military organization, on the one hand, and the eventual societal hierarchy and domination patterns, on the other. From the latter, it is assumed that tribute flows demanded administrative functions, which, along with military needs, led to the eventual formation of a centralized government. According to this external conflict model, then, the city-state arose from conflict over limited agricultural resources. Here the connection between cities and the natural resources needed by a growing population is implicit.

Maisels, using Diakonnoff's (1969) internal conflict model, illustrates how internal dynamics within the growing villages of hydromorphic zones could have accelerated and refined the formation of the city-state. In Maisel's model, three factors are essential for understanding the sociopolitical evolution of cities: increasingly differentiated crafts (i.e., labor specialization), the need for more slaves (i.e., cheaper or free labor), and increased irrigation. These three factors are assumed to have generated increased wealth that led to further differentiation of the poor and the rich. A primitive form of preindustrial capitalism appears to be implied here. Given the increased level of warfare over agriculture and the growing class conflicts, this model suggests that, to a large degree, the birth of city-states was a function of various internal conflicts. However, even a city-state spawned by internal conflicts would focus on irrigation (among the three input factors in this model) as a way to assure productivity and control of civil matters. Nature is once again implicitly identified as a major raison d'être of city-states, but so are sociopolitical issues, especially managing (or eliminating) internal conflicts, albeit in favor of the elites.

Wittfogel (1957) offered a less conflict driven and more managerial model of city-state development. This concept has been used more regularly in various disciplines and forms the basis of many texts on urban history. His model began with large-scale irrigation, assuming that such an agricultural system would necessitate water scheduling, construction planning, labor coordination, and stable productivity. The first factor would lead to the formation of specially structured calendars; the remaining factors would lead to increased wealth. Both outcomes, along with the need to defend irrigation networks, as well as planning and labor-coordination services, would give birth to differentiated and professionalized leadership in city-states. It should be noted, however, that implicit in this model is a structure of class differentiation that leads to various forms of internal conflict as well as to inherent conflicts over natural resources. It would be useful, therefore, to entertain the idea that, while city-states were born from agricultural innovation, conflict over the use of natural resources, which, along with the further separation of social classes (through differential levels of accumulation), were partially responsible for many external and internal conflicts, have accompanied the history of urbanization. Mumford (1961), in fact, connected the rise of cities to the institution of kingship, which became pivotal in establishing authority and organizing the war machine.

These deductive/inductive theoretical models are driven largely by archeological evidence, which, in addition to texts/tablets, includes urban forms as evidence about how ancient societies organized themselves. The morphological pattern of the early Mesopotamian cities provides some of the most intriguing clues about how the first city-states operated and how their design may be connected to their social, economic, and cultural dynamics.

With little variation, the cities in Mesopotamia (as well as in the Indus Valley) had five morphological components: an encompassing wall, a citadel, one or more temples, a granary or food storage, and various residential neighborhoods. The market area was in some cases inside the city but was often relegated to temporary spaces outside the wall. The city, as a place of commerce, had yet to cultivate the full development of market functions. As Mumford (1961) argues, the city belonged to the warrior kings, who, with the help of the temples, assumed control of the city. The five morphological elements symbolize the urban functions of kingship, religion, defense, and urban management. But what was defended or protected? Herodotus's account of how the city of Ecabatan (or Hagmatana) was built by the Medes may provide an answer. He suggests that after being elevated from the status of a just and wise man to kingship, Deioces (seventh century B.C.), the first king of Medes, had a palace built for him that was protected by walls. Medes built strong circular walls, one inside another, to protect not themselves but their ruling elites. This pattern is also visible in the case of Khafajah (from the early dynastic period). In this city, the enclosure of temples and palaces within their own walls suggests that cities may have been built to provide protection, but they protected the ruling and religious elites more than any other group. Thus we should not equate the early cities with the "city of people," which appeared many centuries later. In fact, the morphology of earlier cities needs to be studied separately from those of Greeks and Persians, who inherited more than five millennia of urban experience by the time they built their cities.

The majority of the early cities were small in size and housed very few people. For example, during the Uruk period, the north Syrian town of Habuba Kabira, which is located along the Euphrates, had an enclosed area of eighteen hectares, or about forty-four acres (Crawford 2004). There were only two gates, and the city seems to have had a dominant acropolis that included some residential and production areas (i.e., cottage industry). In Ur the acropolis had its own walls owing to the size that the city had achieved (100 hectares, or about 250 acres) by the time of the third Ur dynasty. While the first layer of walls defined the boundary of the city, it was the stronger inner walls that separated the elites from the others. These walls defended against both internal and external conflicts but, more important, they distinguished "inside" and "outside." Ironically, despite its larger size and clear representation of the five morphological components listed above, public space and major markets were barely accommodated in Ur. As in smaller cities, it appears that these functions occurred closer to the gates and at times extended beyond them to areas outside the wall, but along the road leading to the gate. This pattern remained common in many Middle Eastern cities until the twentieth century (Modarres 2005).

Despite the exclusion of the market, or the inability to give a large space for its public functions, by the third millennium B.C. expanded public works, especially the building of stronger walls and paved streets, became a major feature of cities (Crawford 2004). Water and food storage areas, however, remained within the inner walled area of the city (as in Nippur), and the residential areas, which were located between the inner and the outer walls, were segregated by social and professional status. Many cities today display similar spatial patterns of wealth. However, what distinguishes the geography of class in ancient cities from that in modern ones is that there is little evidence that a land value structure was used as a mechanism for segregation. In fact, the selling and buying of land or houses does not appear to have been a common practice. Families were assigned land by their social status, and the size of the assigned plot was a function of position within the social hierarchy. Many of the permanent labor classes exchanged their labor for subsistence provisions that included being housed by their employers (or owners, in the case of slaves) (Pollock 1999).

Morphologically, not all Mesopotamian cities were alike. For example, the city of Eridu was built with more stone (as opposed to mud bricks) than other cities of its time, was located away from the Euphrates (but closer to where the waters of the Persian Gulf once reached), and had little resident population by the first millennium B.C. One of the oldest cities in Mesopotamia, Eridu was also built around temples. Its origin was associated with religion and its demise in 600 B.C. was caused by environmental factors (namely, the accumulation of silt and the increased salinization of agricultural land in the southern regions of Mesopotamia) and political shifts to the north. By the time Hammurabi ruled Babylon, the Sumerian world moved at a slower pace, and the rise of the Assyrians meant that a complete shift of power to the north had been accomplished.

In both Babylon and Nineveh, urban structures were fully formalized. This included not only the presence of walls, temples, and palaces, but also, as Pollock (1999) has argued, inequalities and sociospatial differentiation. Urban structures included monumental architecture, dedicated to gods and god-kings, and (at least by the time of the Assyrians) a system of interurban connectivity. The emerging hegemony of power in Mesopotamia, a millennium before the arrival of the Persian Empire to the east and the Greek to the north, meant that Assyrians would successfully capture the long-isolated Egyptian cities. Pharaohs, who had enjoyed the natural protections of deserts and the long distances between them and other powerful empires, built their cities without walls. In fact, Egyptian cities were mostly cities of the elite. As Mumford argued, unlike Mesopotamian cities, which grew organically, Egyptian cities were in many cases a one-stage operation. A geometric plan on a largely flat piece of land was followed, which allowed for a quick construction. In more exaggerated form, Egyptian cities followed Mesopotamian cities in their dedication to gods and god-kings. In both geographies, the separation of rulers from the religious authorities was unnecessary and unwarranted.

While planned (i.e., orthogonal) cities were a rarity in Mesopotamia, by the time of the third dynasty of Ur, at least, we find major streets to which many residential alleyways led. The main roads were typically wide and in many cases paved. The walls, symbols of power in Mesopotamian cities, could attain a substantial size, though they were not completely impervious to invading armies (even in the case of Uruk, in earlier periods, the wall was several meters thick and at least seven meters tall).

The absence of geometric or orthogonal cities in Mesopotamia does not necessarily mean that the Greeks invented this concept. In fact, as suggested earlier, Egyptians employed strict geometry for their cities, as did the Chinese. The first urban grid pattern, however, may have been employed in the city of Mohenjo Daro in the Indus Valley (from the third to the second millennium B.C.). This was one of the major cities of the Harrapan civilization and exemplifies an ancient planned city. Not unlike Mesopotamian cities, however, it was divided into two parts. The high city, or citadel, which housed the elites and granary and provided better urban amenities, was located next to the lower city, housing a population of up to thirty-five thousand. The lower city was planned in a grid pattern, with specific attention to water drainage. Though the Harrapan culture and its associated network of cities collapsed before the end of the second millennium B.C., it is important to note that these cities were not completely isolated and were known to Mesopotamians and others who may have established trade with them.

GREEK CITIES AND URBAN DESIGN

To the Western world, and especially to urban historians, Greece is viewed as the source of two major urban inspirations—the rise of a "city of people" and some basic urban planning principles. While neither was invented by the Greeks out of thin air, the Greeks were perhaps the origin of various documented urban concepts that were inherited by Etruscans, Romans, and, later, Europeans. The idea of the city as a place of freedom and social growth was strong among the intellectual elites of Greece. For example, while Plato was cognizant of the inherent conflict within and between cities, for Aristotle the city was a place where people came to live, and where they remained in order to live better. It is interesting that as cities of people rather than as places dedicated to god-kings alone, Greek cities required an imposed order that, despite its expressed language of dedication to divinities, was invented by people.

We look to Hippodamus as the father of Greek urban planning, but before we begin a discussion of the cities he designed and his influence on the Etruscans and Romans, we need to look at Greece before the Persian attack that led to the destruction and reconstruction of Miletus. About the same time that Deioces had the Medes build him the city of Ecbatana (as Herdotous tells us), what is known as ancient Greek civilization emerged. Derived from three earlier civilizations—the Minoan, the Cycladic, and the Hillidic (which flourished along the Aegean Sea beginning in the third millennium B.C.)—Greek civilization, as we know it today, appeared by the eighth century B.C. By the sixth century B.C. Pythagorean geometry and its search for a relationship between spatial and musical proportions had led to various exercises in architectural aesthetics and, by extension, the urban form.

But fully realized rational planning did not begin until the building of Miletus. In Athens, for example, we cannot detect a formal plan until the fifth century B.C. However, as Bacon (1974) illustrated, the Panathenaic Way, which functioned as a local and interregional thoroughfare, acted as a spine around which various morphological components of Athens were located. Athens's agora (marketplace) and acropolis were connected by its pathway. Here, the street became both a sacred way, connecting various temples, and a commercial and public route. During the various stages of its development, Athens grew organically around this major thoroughfare, and its agora evolved with the gradual growth of the city itself. What made the urban structure of Athens interesting was its use of the shaft of space created by the Panathenaic Way. Even by the second century A.D., Athens's agora was fully adapted to this street.

What we recognize as Greek city planning comes from the rebuilding of Miletus after it was destroyed by the invading Persian army. In reconstructing the city, Hippodamus formalized the Greek urban morphological components (i.e., agora, acropolis, and gymnasium). (Theatre was added later—in the case of Miletus, during the Hellenistic period, in the middle of the second century B.C.) Hippodamus also introduced the concept of orthogonal (or gridiron) planning through repetitive modules of rectangular blocks. His geometric layout remained in use (and in fact the blocks were gradually filled) until Roman times. Miletus thus accommodated a historical progression of change within its preconceived geometry from the fourth century B.C. to the second century A.D.

Miletus is important not only as the place where the gridiron plan was first implemented on a citywide scale but also for the promotion of the Hippodamian plan. The city of Delos, located on the island of Deloa in the Aegean Sea, also followed an orderly plan. Different parts of the city, though separated by topographic features, fit well into each other and the surrounding environment. In fact, Delos is a great example of how orthogonal plans can be incorporated into natural settings with minimal need for total leveling and destruction of the site.

The art of Greek urban planning was displayed at its best in Priene, a Greek colony in Asia Minor. The city was built in a gridiron pattern, with attention to movement, elevation, and placement of various morphological components in an awe-inspiring spatial arrangement. Priene is not only a great example of the merging of architectural concepts with urban design; it is also a place where the connection between the Greek view of order and social structure is clearly embodied in the physical layout of the city. The search for order did not diminish the Greeks' attention to gods, which was evident in the placement of temples on the highest grounds or other visible locations. The Greeks' concepts of democracy and social engagement directly informed the building of their cities, for a rational plan was seen as the ultimate expression of democratic order (Bacon 1974) without unnecessarily diminishing the importance of the gods. Ironically, the Greeks built few cities, other than Miletus, in Greece that benefited from Hippodomian concepts in urban planning. In fact, they were more likely to export their orderly urban design concepts to their colonies in Asia Minor and other Mediterranean regions, as opportunities to build new cities were greater in those areas than in Greece.

In addition to Asia Minor, the Greeks built cities in Egypt (e.g., Alexandria) and Syria (e.g., Antioch). But it was their orthogonal planned cities in southern Italy that had the most profound impact on cities in the West. Paestum, Agrigento, Naples, Heraclea, Tarentum, Megara Hyblaea, Piraeus, and Thurii were among the cities built or rebuilt using orthogonal Greek planning guidelines. Etruscans, whose cities may have inspired the Roman emperors, followed this planning process as well. Some historians, including Anderson (1997), remain unconvinced that the Romans learned their town planning from the Etruscans, but it seems clear that Greek colonial cities in southern Italy may have directly influenced the Romans.

ROMAN CITIES: URBAN PLANNING AND MANAGEMENT

Greek urban planning concepts and morphological elements appeared in Roman design in an elaborate and purposeful way. In addition to the planning influence, which is assumed to have come from Greek colonial cities in southern Italy, the intellectual lineage in articulating urban planning concepts is also important. From Hippodamus to Aristotle, theories about an ideal city, including its shape and size, were textualized. Greek ideas, then, were carried not only through the physical examples of their gridiron cities but also through the transference of these formal concepts, which appear in texts by Dodorus and Strabo. By the time Romans began to build their new towns—which included military camps, called *castra*—the logic of orthogonal planning was fully documented and made available to the builders of the empire. Vitruvius, whose directives are evidence of Roman urban planning style, was one such inheritor of the Greek intellectual heritage. But Vitruvius's orders were not implemented exactly; his concepts were simply the ideals of planning. Just as we today do not follow our normative planning rules in building cities, we should not expect Romans to have followed Vitruvius's guidelines precisely. We shall discuss Vitruvius's ideas later; for now it is enough to note that, as was the case with Greece, Roman standardized planning occurred more rigorously in the empire's occupied territories than in Rome and cities closer to the seat of the power.

As with the Hellenistic phase of Greek civilization, when Greek ideals, aesthetics, and fashions spread throughout the ancient world and were a sign of being "cultured," the Romanization of the Roman Empire's subjects was an important step in tightening its grip on conquered lands and peoples. As Tacitus said of the Roman influence in Britain during the first century A.D., not only did the Britons adopt the language and clothing fashions of their conquerors, they also "were seduced into alluring [Roman] vices: to the lounge, the bath, the well-appointed dinner table. The simple natives gave the name of 'culture' to this factor of their slavery" (Anderson 1997, 183–84).

In addition to Romanizing their subjects through language and material culture, Romans recognized the importance of imposing order through managed urbanization. The visualized order granted by orthogonal planning and organized urban management techniques became necessary tools for maintaining a vast empire that spanned an area from Britain to North Africa. For this reason, the invention of urban classification, micro- and macromanagement techniques, building codes, and many other concepts were the Roman intellectual contributions to the urbanization process.

While the Greek and Etruscan empires were federations of city-states, Romans incorporated all cities and settlements within a hierarchy that formed their empire. The state determined the fate of a city, and the idea of a totally independent city was largely nonexistent. In a centralized empire tributes were necessary, and cities were part of a large financial network that had to be managed in a formal manner. For this reason Romans divided their cities into three distinct categories—*colonia, municipium,* and *civitas*. Colonia was the highest ranking a city could receive, implying that the majority of its residents were Roman citizens. A colonia was governed by a council under the supervision of four magistrates, appointed annually. In Britain and other colonies, cities in this classification category were not numerous and typically housed retired legionaries, who helped reproduce the state culturally and biologically in the far corners of the empire.

As opposed to colonia, a municipium (the second-highest-ranking city) was home to a smaller number of Roman citizens. Such a city would be governed by a council under the direction of four elected magistrates, who at the end of their service were awarded Roman citizenship. Direct engagement in governance, which legitimized the state, was awarded with cultural, social, and political Roman identity.

The last and the lowest urban rank was given to tribal towns. A civitas lacked any Roman sociopolitical and cultural affiliation, since it housed very few Roman citizens. Under a Roman provincial administrator, these cities were managed as independent administrative centers governed by the local population. This does not mean that a civitas had no council or magistrates, but rather that the local aristocracy participated in managing local affairs, becoming Romanized at a slower pace than the other two kinds of cities.

These three urban classifications, which were fully implemented in Britain, provide a clue into the Roman urban management genius that attempted to create the hegemony of *Romanitas* within its conquered territories. The urban form introduced into Britain and other colonized areas was simply a continuation of this managerial and state-building project. From Bath to Gloucester, Romans wove their management techniques into urban design. The result was a lasting urban legacy that continued through the medieval period. In order to learn about their planning approach, we turn to Vitruvius and a few Roman cities.

While a complete treatment of Vitruvius's planning guidelines are beyond the scope of this book, the basic elements of his treatise are important for understanding the morphology of planned Roman cities (as well as for understanding what a "city" meant to Roman elites). Vitruvius divided urban morphological elements into two groups: city walls and public buildings. He avoided any discussion of residential, or what may be called vernacular, architecture, with the exception of discussing the design of elite housing units. Vitruvius thus became a voice of the urban managerial elites, whose conception of how a city should be built and managed was important to the empire. For Vitruvius, the wall was both a demarcation of the urban boundary and a symbol of defense. The gates and towers were one of his three categories of public buildings. The other two categories were religious (e.g., temples) and public service (e.g., forum, harbor, and colonnade) buildings and spaces. In his theory, urban planning began with selecting the site and determining where the wall should be built. After that, streets (or transportation network) were designed, with special consideration of the prevailing wind in order to protect alleyways and streets from the wind. In fact, Vitruvius's comments suggest that the Romans were concerned about the basic health of urban residents and sought protection against harsh climates. Once the basic transportation backbone was fully designed, religious buildings were sited. Vitruvius provided a list of religious and public buildings, ranked by their order of importance. The ideal Roman city followed a formulaic sequence.

The spatial structure of a planned Roman city was much the same as that of a Greek city. The cardinal direction of the city was defined by its major transportation routes; the residential and commercial sections were arranged along these routes. But what gives each city its distinguishing features were the size and location of public and religious buildings. Romans extended the Greek agora and acropolis into their cities, but their forum and amphitheatre (which were included more regularly during and after the Augustan age) were built on a different scale and at times for a different social agenda. As we shall see, Romans also added an element of verticality to their cities. New construction materials and techniques allowed them to add to the height of their buildings, permitting a greater population density. This meant that building codes, public services, and urban management concepts had to be refined in order to manage the larger cities.

Despite the apparent structure introduced into the empire and its far-flung cities, the seat of power—Rome—received no serious planning or spatial reorganization until the time of Julius Caesar. His architectural contributions were monumental and gave the city its first imperial look. For Rome, however, gridiron planning was impossible. The city had to be organized through planning and management concepts rather than by physical layout alone. This does not mean that Rome lacked monumentality. In fact, architects of Rome used new buildings to give the appearance of rational organization to a city that had existed for many centuries before the rise of the Roman Empire. As Bacon (1974) explains, the sheer mass of interrelated buildings, their scale and visual containment, gave the city its aesthetic organization. To this was added movement through the Triumphal Procession of Rome, which, unlike the Panathenaic Way in Athens (which extended the length of the city), coiled into a single space. The Circus Maximus became the most important site of this symbolic movement, transferring the need for linearity of space across a city into the circularity of a single building. A mass of individually functioning buildings and their interrelationship created the planned connectivity that could not have otherwise been achieved within the organic shape of Rome. Beyond these architectural innovations, however, order and regularity of space had to be achieved through means other than gridiron planning. The sheer size of the city (in area and population) necessitated innovative management techniques and regulation of spatial functions. We look at Rome as the site of the first Western experience with complex urban governance.

From the foundation of the Roman Republic in 509 B.C. (when the Etruscan monarchy was overthrown) to 400 B.C., Rome grew minimally, from a population of twenty-seven thousand to twenty-nine thousand (Ward 1990). Spatially, however, the city grew from 830 to 1,050 acres. This translates to a small decline in population density of the city of Rome in the first century of the republic. By the first century B.C., however, Rome had grown into a major urban complex. Estimates of the population during the Augustan age, which lasted into the first century A.D., often exceed 1 million.[1] Even if population figures were smaller,

---

[1]For a discussion of Roman demography and the population size of Rome, see Wiseman (1969); Packer (1969); Ward (1990); Patterson (1992); Casico (1994).

other evidence from the era indicates the enormous size of this city, a size not achieved by any subsequent city until the arrival of modern industrial cities. For example, the presence of more than forty-five thousand *insulae* (i.e., apartments) in Rome points to a very densely populated city (Packer 1967). It is therefore not surprising that Julius Caesar and Augustus established an urban agenda that rebuilt and reorganized the city. Without a doubt, these rulers were responding to the needs of the empire and its embodiment in the capital city. But it should be remembered that running a city of roughly a million people was no easy matter for a society that lacked sophisticated communication techniques. Distance had to be overcome through spatial arrangements and rules of movement that would remake the spatial structure of the city and its morphology, even before the fire of A.D. 64.

Caesar's plan for remodeling the city shocked many of his contemporaries, including Cicero, a political philosopher and commentator of the first century B.C. (Anderson 1997). He added the Forum Julium, Basilica Julia, and Circus Flaminius (which was finished as the theatre of Marcellus). Many of these monumental constructions reoriented the city both morphologically and functionally. Caesar even reworked certain sections of Campus Martius, located in the northwestern section of the city. After the assassination of Julius Caesar in 44 B.C., Augustus continued the rebuilding of Rome in earnest. He built a number of temples, the first Pantheon, public baths, and the Basilica Neptune. He also ordered restorations throughout the city. The city was transformed into an urban complex of close to a million people, living in an impressively large number of apartments stacked vertically and located along a labyrinth of street networks. This everyday architecture was overshadowed, however, by a mass of monumental architecture whose visual imposition gave the city an air of order.

Rome's urban components can be divided into six groups: the wall, street system, housing, markets, city centers, and recreation. The first wall was built in the sixth century B.C. and remained operational until the first century B.C. owing to Rome's smaller size. With the rise of the Roman Empire, Rome had little use for a wall. As the seat of power, the city was in little danger of invasion. Spatial growth beyond the older wall thus became possible. By the end of the third century A.D., however, a wall had become necessary. The Aurelian walls were built at the end of that century, encompassing the spatial extent of the city after nearly three centuries of growth. This wall was eleven and a half miles long, encircling an area slightly larger than five square miles. This made for a city with a high population density and crowded streets.

Roman cities usually had three major kinds of streets: *itinera, actus,* and *viae.* Itinera accommodated people on foot; actus allowed for the passage of one cart and pedestrians; and viae provided enough room for the passage of two carts. Given the crowded nature of Rome and the fact that viae could not be numerous, the first traffic laws were introduced in the first century B.C. One of these regulations limited the movement of carts during the day, especially if the cart was used to transfer goods. This minimized the mortal danger facing pedestrians who had to encounter rapidly moving carts on crowded streets.

Housing in Rome fell into two categories, *domus* and *insulae.* The first consisted of single-family structures (with multiple design possibilities, which Vitruvius discussed), and the second of apartments. Domus were typically occupied by the upper classes and insulae by everyone else. The worsening of the social class division in Rome can be seen in the number of housing units in each category. By the fourth century A.D., Rome had in excess of forty-five thousand insulae and fewer than two thousand domus (Packer 1967). Given the size of the city, such a large number of housing units had to be stacked up vertically. Although the practice of building to two or three stories had been established by the third century B.C., it became important in the first century B.C. to limit the height of buildings. Building codes were thus enacted as early as the reign of Julius Caesar, and at various times building heights were limited to sixty or seventy feet. These regulations tempted Roman builders to create common walls between various structures, as this allowed them to use every square inch of land available. Given that wood was an important construction material, it became necessary to consider the risk of fire as well. The minimum distance between buildings and the requirement for using fire resistant tiles for roofing were legislated by the government of Julius Caesar. This did not solve the fire problem, however, and Augustus formally created a group of firefighters to respond to this growing urban problem.

To Augustus, Rome appeared in need of efficient management. More important than establishing fire-fighting groups was his idea of dividing the city into fourteen regions, with ward divisions and management teams. The layout of the regions suggests a logical spatial division that assigned to each region its share of public buildings and public space to manage. This creative localization of urban services and management through selected magistrates was effectively implemented, and its basic ideas have been inherited by many cities in the West. Innovations in urban management went beyond this localization of governance and included the formation of policing functions, sewer services, and so on. The Augustan age was truly the age of urban revolution, not only in its impact on Rome but in the diffusion of ideas about cities beyond Rome.

The city center can be counted as the fifth morphological element of Rome. The center was located between the hills of Palatine and Capitoline and the end of Quirinal ridge. It grew in two ways, both through controlled growth toward what eventually became the Colosseum (originally called the Flavian amphitheatre) and through successive planned *fora*, which were built from the time of Julius and Augustus down to the beginning of the second century A.D. With the exception of the Forum Julium, the majority of the planned fora were built to the north of the Colosseum.

Markets were critical to Rome, especially given that the city depended almost entirely on imports. This necessitated the establishment and maintenance of highly developed ports and warehouses. From these locations goods had to be carried to individual shops, many of which occupied the first floor of residential buildings. The pattern of housing being located above shops can still be found in many regions in Europe and the United States. In addition to residential areas, shops could also be found around fora, especially the Forum of Trajan.

In addition to the market and the fora, a large urban complex required a significant investment in entertainment, the sixth morphological element. This need was especially accentuated by high rates of unemployment and jobs dependent on public works. Entertainment was a form of citizen pacification. The enormous size of entertainment buildings, such as the Colosseum and Circus Maximus, which could accommodate more than two hundred thousand people, suggests the Roman economic situation of massive wealth accumulation beginning in the first century, and drastic worsening by the third century A.D. Roman holidays also increased in number, allowing employment for a larger number of people. In this case, the urban morphology of Rome, and for that matter of many cities in the empire, reflects the social, economic, and cultural conditions in the empire.

Despite its eventual decline, Rome can be seen as the birthplace of the Western urban ideal, where planning, management, and architecture converged to create an image of order and hegemony. Though

Romans implemented the gridiron pattern more frequently in other cities, in Rome they enacted the urban policies and impressive public works programs that became their true urban legacy for the West. Medieval European cities inherited the Roman urban system, and despite failing to maintain continental interconnectivity, in many cases they built upon a Roman infrastructure and governance structure. Even today one can see the influence of Roman concepts in Western city council districts, geographically structured policing, fire departments, and other urban services. While Rome was in many ways an exception to the urban model of its time, and even of the urban planning principles Romans espoused, it is truly the birthplace of civil society and of Western concepts of urbanization.

EUROPEAN CITIES IN THE MEDIEVAL PERIOD

The fall of the Roman Empire and the disappearance of a formal governing structure established by a central government brought about a return to city-states and a localized (i.e., feudal) urban system. In a millennium that ended with the Renaissance in the late fifteenth century, Europe went through significant waves of social and cultural transformations that have erroneously been called the "Dark Ages." In fact, there was nothing "dark" about this era. It is thanks to Renaissance thinkers that we employ this disparaging terminology even today. While the Muslim empire flourished from the seventh to the fourteenth century, prospering through commerce and intellectual achievement, we are told that overzealous religious establishments and abusive feudal lords subjugated the good people of Europe to a life of ignorance, stagnation, and poverty. While there is no denying that the social system was abusive toward its lower classes, this is hardly an adequate explanation for the demise of cities. Neither Romans nor their appointed rulers were any kinder than the medieval feudal lords. The secular cities that arose after the medieval period were no easier for the lower classes to live in, either. The dark images of abject poverty, epidemic disease, religious demagoguery, witch hunting, interregional wars, the Crusades, and many other negative features of the so-called Dark Ages may be more closely related to the disappearances of large-scale intercontinental commerce (or to our skewed perceptions of premodern Europe, inherited from the Renaissance) than to objective fact. Being disconnected from Asia (albeit through the familiar route, which extended from Anatolia to China) by the geographic presence of the Muslim empire meant that European urban economies became more local. Nonetheless, it would be a mistake to assume that European cities ceased to exist or lost their importance. In fact, the urban morphology of the European medieval cities provides an alternative history of this period.

Though cities like Rome did not emerge for many centuries, by A.D. 1200 there were about 250 cities in the Holy Roman Empire, west of the Elbe and the Saale (Dickinson 1945). While the number of towns was significantly smaller to the east, the growth of Germanic towns to the west points to a continuing urban tradition in western Europe. During the four centuries preceding the Renaissance, thousands of new towns were founded in the Germanic cultural area alone. Medieval Europe, especially from the thirteenth to the fifteenth century, witnessed a remarkable growth manifested in a rapid rate of urban expansion.

Most medieval European cities had small populations (fewer than twenty-thousand people), and their morphological pattern was a function of their historical origins (e.g., Roman *castrum* in the case of cities like Regensburg, along the Danube in Bavaria). While the Romanesque period (circa A.D. 1050–1200) witnessed a growing presence of the church, both sociopolitically and architecturally, the Gothic period (roughly 1200 to the Renaissance) saw an acceleration of this process, along with the gradual expansion of commercial activities. As cities began to expand demographically and structurally, the growing architectural presence of the church and the powerful elites provided an important context for the urban forms of the late medieval period.

Spatially, medieval cities were compact and contained a number of recognizable elements. These were fortress, charter, wall, market, church/cathedral, roads, and residential/vernacular architecture. The fortress was an important symbol of safety and protection, which in turn attracted a residential population. More important, however, was the charter, which allowed city dwellers to become free citizens. The medieval city thus became a city more of people than of serfs.

By design and by the nature of transportation technology, there were very few wide streets in medieval Europe. Conzen (1960) identified three functional types of street in medieval cities, from England to the planned *bastides* of France. These were heavily trafficked streets, residential streets, and occupational roads. Stores and residential units were in many cases stacked vertically along residential roads, a pattern observed earlier in Rome. The gradual agglomeration of market functions along major roads and plazas appeared by the late medieval period. Many of these urban elements are still present within the spatial fabric of older

European cities. Medieval urban morphology and its architectural heritage are thus relevant to understanding some contemporary cities, as well as the daily lives of their residents, during this historical period. A significant portion of these medieval cities remain at the core of our modern urban fabric, and, while providing marketing opportunities for tourism, they are hindrances to mobility, access to modern urban services, and urban growth. In major cities, from York to Vienna, older sections are protected for their economic and cultural values (e.g., in some cities, automobile-free zones have been created), but in smaller cities adequate resources for their protection are unavailable. In this the medieval sections of European cities share a common predicament with the old *medina* of North African and Middle Eastern cities. In fact, Asian cities such as Kashgar in western China face similar problems. Unlike many non-European cities, however, the medieval towns have fared relatively better in architectural preservation and spatial accommodation of modern activities.

Beyond providing protection, medieval city walls established a social, cultural, economic, and political boundary between city and country. The importance of this boundary is evident in the common practice of distinguishing city residents from nonresidents. In some cases nonresidents would be asked to leave the city at sunset, which led to the creation of encampments outside the city walls. Occasionally, when these temporary arrangements grew in size, cities would expand their walls to incorporate them.[2]

If walls and fortifications provided protection for city dwellers, it was through commerce that living in towns became more viable. Urban growth was based largely on the expansion of regional trade, especially in textiles. From Italy to England, cities with growing textile production, distribution, and retailing witnessed significant growth. Businesses related to textiles, as well as urban services such as inns, which accommodated traveling merchants and their staff, also gained economic momentum. From Venice to Brugges, Amsterdam to London, this growing urban market meant that more sophisticated ports and warehousing developed, and that a growing portion of the population could be employed. In fact, urban growth during the medieval period fed and stimulated, in a reciprocal manner, the agricultural economy of rural areas as well. The centrality of the textile industry meant that a decline

[2]Expansion of walls to incorporate adjacent settlements and small villages has been documented in a number of publications on medieval cities. For example, see Nicholas (1987).

in the market could also be hazardous to the survival of a city and its hinterland, as the case of Ghent may illustrate.

Given the small populations of medieval cities, it is natural that market areas would require little land (Venice is a notable exception). As regional commerce began to grow, however, market space not only expanded but occupied areas adjacent to the gates and along the main roads. Many of the occupational roads were dedicated to this function. Fairs and growing markets at times spilled over the walls and the city gates to occupy spaces outside the city limit. This morphological evolution was also common in many cities in the Middle East and eventually determined the direction of urban growth in its next cycles of expansion.

There is no escaping the fact that the church eventually became a pronounced architectural presence in cities. From the Romanesque to the Gothic, the church gradually moved from a marginal to a central position of power in the medieval European city. Cities have rarely been without a significant religious presence, and religious buildings have always been more than places of worship. They were and continue to be centers of social and civic interaction as well. The alliance between the religious establishment, the landed gentry, and the governing body, then, was hardly new to the medieval period.

Despite the growing significance of the church, medieval cities put new emphasis on the freedom of urban residents, as evidenced by city charters that enabled them to live as free people. Vance (1990) promotes the image of the medieval city as a city of people. He argues that urban form in medieval cities responded more clearly to citizens than to gods. This could also be related to the growing number of artisans and merchants who formed a new urban bourgeoisie that helped stabilize the social, economic, and political structures of the city. Contrary to received wisdom, medieval cities were not awakened from a slumber by the Renaissance but rather gave birth to the Renaissance.

Medieval cities regained symbolic importance during the eighteenth and nineteenth centuries, when European antimodernists and Romantics began to look upon them nostalgically, whether as environmentally adaptive cities or as organic urban forms that elevated the sense of community among their residents. From Wordsworth to Camilo Sitte, the organic form of the medieval city and assumptions regarding its "nature" were used to promote alternatives to modernity and its urban forms. The Gothic revival of the nineteenth century was also a nostalgic architectural return to an era preceding the neo-Platonic world of the Renaissance and the Enlightenment.

MIDDLE EASTERN CITIES: A MORPHOLOGICAL SYNOPSIS

Space constraints prevent an extensive discussion of the morphology of Middle Eastern cities, but we cannot understand urban morphology without at least a brief look at the Middle East.

Many cities in the Middle East—the large region extending from Iran to Lebanon and Turkey to Yemen—are built upon or located near the site of Mesopotamian, Median, Harrapan, Hittite, and Persian cities. It should come as no surprise that the premodern Middle Eastern urban form, at least in the old medina, is similar to that of older Mesopotamian cities, and for that matter to the medieval cities of Europe. Any compact city that responded to topography, environmental factors, and other situational variables would produce similar spatial patterns. While the similarities between Imola in Italy and Herat in Afghanistan are striking, urban morphology is affected not only by social and cultural factors but also by commerce and, even more important, available building technology. Many Middle Eastern cities were achievements in their own right, even though they may have been part of a large empire. Given their level of insularity and self-dependence, they all had walls, fortifications, major roads, markets, and a citadel that housed local ruling elites and their staff (especially when rulers were appointed outsiders).

We take issue with the older view that religious (i.e., Islamic) factors explain urbanism and urban patterns in the Middle East. William Marcais's 1928 article "L'Islamisme et la vie urbaine" (Abu-Lughod, 1993)[3] influenced a number of scholars, who took his general concepts about Islam and its impact on Middle Eastern urbanism, which was based primarily on his familiarity with North African cities, as gospel.[4] By contrast, we follow Abu-Lughod's argument that in order to understand the Muslim contribution to Middle Eastern urbanization, we need to pay special attention to the dynamics of juridical classes and their turfs, gender segregation, and the social construction of the neighborhood as the building block of civil society. Not unlike the Roman, the Muslim empire introduced new organizing concepts that affected urban physical structure and social dynamics, but Muslims had few ideas about how to design a city. This does not mean that Middle Eastern architects never developed a special architectural style for individual buildings. If one were to travel from Morocco to western China, however, it would

[3]Marcais (1928). Marcais's article is discussed in Abu-Lughod (1993).
[4]For example, see von Grunebaum (1955).

become clear that there are distinct variations in design from one region to the next.

The importance of Islam as a new paradigm in governance, commerce, and civil society, as opposed to religious decrees for urban design, is evident in the fact it took nearly three centuries before a significant Muslim urban society could be observed. It was not until around the eleventh century A.D. that most urban inhabitants in the Middle East became Muslim. Also, in most cases, the seat of power took little interest in urban affairs. This means that local government, in many cases, was the only government recognized by the locals, and often the local ruler was a native of the city.

Islam's contribution to the emergence of the flourishing urban world in its domain was partially due to the following factors:

- literacy (i.e., production of books, building of libraries and learning centers; construction of *Madreseh*—the city as a place of knowledge);
- the centrality of community at all levels, building a civil society from the ground up;
- the need for consultation (e.g., Caliph as spiritual and political leader without legislative authority in the early phase of Islam);
- development based on spirituality and practicality (commerce was not frowned upon but usury was; emphasis on earning through production rather than investment without labor);
- racial, regional, and class equality; and,
- traveling, which was promoted as a means of learning.

Clearly, these underlying principles encouraged a literate society of like-minded urbane individuals who engaged in commerce and had an artisan lifestyle. Between the eleventh and twelfth centuries, when urbanism achieved a significant momentum, cities from Central Asia to Spain housed a flourishing economy and rapidly advancing scientific communities. The extensive geography of the empire meant the domination of major global trade routes and a significant market expansion. Cities, as sites of commerce, flourished and gained relative importance based on their level of economic productivity. Naturally, prosperous cities like Bukhara, Samarkand, Aleppo, Baghdad, and Cairo also attracted a large number of literati from every corner of the empire. The cosmopolitan state of Baghdad in the twelfth and the thirteenth centuries rivaled that of the great cities of the present day. Like Aleppo and Damascus, the city was home to learned individuals who translated and debated the work of classical Greeks and developed treatises on philosophy, chemistry, medicine, astronomy, and other sciences.

By the fourteenth century, when Ibn-Khaldun and Ibn-Batuta traveled extensively in the empire, the Muslim world they documented was highly advanced and had witnessed the birth and growth of many cities. By then, however, the demise and loss of control in certain localities was becoming obvious. By the time Europeans arrived during the European Renaissance, the Muslim world was in decline, its days of glory behind it. From then until the twentieth century, the region witnessed the rise of warring factions from Central Asia to North Africa. European colonization of various regions of the Middle East in the eighteenth, nineteenth, and twentieth centuries only added to the problems facing an embattled region. Having been occupied for millennia by various cultures, Middle Eastern cities can boast not only a long historical existence but also a hybrid architecture and urban form influenced by various internal and external morphological forces.[5]

Middle Eastern cities were largely produced through site and situational factors familiar to other cities throughout the world. For example, traditional cities in Iran were mainly located on trade routes. What sets them apart is that the Silk and the Spice Roads passed through desert fringes, where water could be found. Despite the pronouncements of Orientalists' assumption regarding the model of the Islamic city, it was water (hydrography and topography) that played an important role in the urban morphology of cities in this region.[6] In many cases, *Qanats* (aqueducts) determined the site and the size of cities in the premodern era.

The morphological elements of the traditional Middle Eastern city were permanent central markets (*bazaars* or *suqs*) with specialized sections (which were functionally isolated, not only for guild purposes but also to ease the taxation process), mosques (including a Friday/congregational Mosque), shrines (especially in Shi'a Iran, where they appear mostly as *Imamzadeh*), public baths, city walls, a citadel (in some cases only after the twelfth century), inns, and smaller markets in individual neighborhoods. In Iran, *Hoseinieh*, where religious passion plays are performed, can also be found. These are small open plazas that today may include parking space for cars. As in the European medieval cities, the older sections of the Middle Eastern cities contain narrow winding alleyways that cannot accommodate most cars. The transformation of

[5]See Al Sayyad (2001) for various case studies of this phenomenon.
[6]For example, see Bonine (1979).

older small plazas into parking lots allows neighborhood residents to park close to home.

Among the morphological elements of Middle Eastern cities, the bazaar plays the most important role, both commercially and politically. Historically, bazaars were so important that when a new one was built the entire core of a city could shift, as was the case in Isfahan, Iran. Despite Orientalists' obsession with the mosque, the bazaar is arguably the main focus of traditional Middle Eastern cities. It is not only the center of commerce but is also the cultural, social, political, and religious lifeline of the city. In Iran the bazaar and the merchant class have acted as the third sector (the others being government and the intellectual elite) whose alliance with the religious establishment and economic ties with the government allowed it to be the sociopolitical conduit between the two groups, at times marginalizing the voice of nationalist intellectuals. As the religious establishment began to make its way into the political world, the bazaar facilitated the process. The Iranian constitutional revolution of 1906 and the recent Islamic revolution of 1978 would not have been possible without the significant role played by the bazaar.

The arrival of modernism spatially disrupted cities from North Africa to western China. Building wide streets through the spatial fabric of traditional cities to accommodate cars and lining the streets with shops affected not only the viability of the bazaar but also the sociocultural organization of Middle Eastern cities—this has been the mark of new urbanism. When one thinks of the current state of Middle Eastern cities, one should imagine a distorted interpretation of Western urbanism forcefully woven into the old spatial fabric of many cities. In mimicking the West, however, Middle Eastern cities have inherited many of its social pathologies as well.[7] From Cairo to Tehran one can observe persistent patterns of pollution, inadequate urban services, traffic jams, suburbanization, segregation, unemployment, housing shortages, and crime. These things are all part of the contemporary urban experience in the Middle East.

RENAISSANCE AND BAROQUE CITIES: RATIONALISM AND URBAN DESIGN

Historical accounts of why Europe witnessed its Renaissance in the second half of the fifteenth century usually single out three factors: the fall

---

[7]Space constraints prohibit a comprehensive discussion of this topic. For more information, we recommend Hooshang and El-Shakhs (1993); Bonine (1997); Elsheshtawy (2004).

of Constantinople and the end of the Byzantine Empire (which is assumed to have brought various books, especially those of the classical Greeks, back to the West), the invention of the printing press (which allowed for proliferation of books), and the historical readiness of the masses for something other than the medieval life of limited commerce, religious domination, and feudal overlords (which left the merchant and artisan classes out of the political power network). Such accounts fail to address a number of issues. First, paper originated in China and was improved and made lighter and thinner (which meant that books were made lighter and more transportable) by Muslims. So, first, why would Europeans necessarily benefit from the invention of the printing press, before they had access to suitable paper for the mass printing of books? Second, what would they publish, and who would buy it? Third, why would Europeans become interested in everything old—why would they develop a love of history—as opposed to focusing on and inventing a future?

Many writers have attempted to answer these questions, but only one applies directly to our case of urban design and morphology. Lesnikowski (1982, 18) suggests that "[t]he saga of the 'pagan' Renaissance begins with the moment when Poggio Bracciolini found some ancient texts in the Swiss monastery of Saint Gall while attending the council of Constance. Among them was the famous treatise by Vitruvius," the Roman urban theorist whose treatises on urban design influenced the Roman elites and their urban agenda for the empire. It is interesting that architecture, urban design, and art became the venue for conceptualizing and reconstructing an imagined golden age of the classics.

In fact, the European Renaissance reached a major turning point when the old texts began to shake the foundation of what was being espoused as the Western tradition in the medieval period. Ironically, the Renaissance was also a reawakening of everything historical and classical—a nostalgic journey back to the individualism and presumed rationalism of the ancients—and a reaction against the stranglehold of religion on public life. Hence Plato became the foundation upon which Renaissance secular-rational philosophy rested; and Vitruvius became the source of treatises on rational urban planning. If "[t]he medieval city found its form by reacting directly to differences in the landscape, allowing the system of streets and buildings to give way to even minor differences or difficulties in the terrain. . . . [t]he Renaissance city interposed between the landscape and the physical form of the buildings an abstract, preconceived ideal based on straight streets and a simple geometrical order" (Kvorning 2002, 117).

Renaissance planning became an exercise in utopian dreams about order—rational spaces for rational people and their societies. Pure geometric proposals, which disregarded topographical features and acted to dominate the "natural" landscape, became a hallmark of Renaissance planning theory and practice. The 1592 ideal city of Buonaiuto Lorini is an example of such aspirations. However, an important aspect of this ordered urban ideal was the impact it may have had on the design of a number of American cities, including Philadelphia.[8] In the Renaissance reconceptualization, nature was also ordered. The green spaces within each city, including those built for the elites, were designed as tortured geometries of nature meeting straight lines.

As Europe began to navigate its age of exploration and to prosper from its expanding global power, the functionality of the medieval period was replaced in European cities with concern about external spaces, aesthetics, and monumentality. The Renaissance was more than an exercise in the aesthetics of individuality. With the rise of strong central governments, larger numbers of elites could take up residence in the countryside, allowing for the separation of classes and the creation of a class-based spatial distinction between cities, suburbs, and countryside. As Vance (1990) has suggested, the Renaissance can be seen as an important phase in the development of the class segregation of modern European cities.

Renaissance urbanism was not all bad, however. From Brunelleschi's dome in Florence to the Piazza della Santissima Annunziata and the Piazza della Signoria, the technological advances and aesthetic innovations of the Renaissance introduced a wholly new scale and visual experience to cities. Linear streets, geometric spaces, and repetitive geometric forms, culminating in monumental architectural forms, produced a visual depth that exuded order and wealth. As Bacon (1974) has argued, however, the Renaissance designers failed to see the "total city problem." The city thus became an extension of exercises in classicist architecture and perspective drawings that formally introduced the concept of movement to architectural design. Renaissance urbanism itself was based primarily on discarding medieval cities and building upon them imagined, ordered places of classical beauty. By the end of the sixteenth century, as the

[8]Vance (1990) argues that the design of Philadelphia has little to do with the Renaissance planning approach (264). However, the orthogonal plan of this city, which is based on long, straight streets and repetitive simple geometric blocks, is an American interpretation of the Renaissance by Bacon (1974, 218). In fact, Savannah's design is also seen as having been similarly influenced.

Renaissance waned, mannerist architects and designers had become fully engaged in designing ideal cities, which in turn informed baroque concepts in urban design, albeit in a theoretically altered format.

By the end of the sixteenth century the European public had become weary of the individualistic Renaissance aesthetic and the social philosophy that informed it (Lesnikowski 1982). Public desire for a centralized order provided the Roman Catholic Church with an opportunity to assume a powerful position in the seventeenth century. With the return of church ascendancy and its dogmatism, cities such as Rome became architectural playgrounds for inventing geometrically ordered space in and around churches. In fact, from the time of Sixtus V, Rome had witnessed a remarkable series of urban renovations under the direction of great masters such as Michelangelo, Bramante, and Carlo Maderna, whose initial project for Saint Peter's dome proved to be structurally problematic. However, Maderna's design for the piazza in front of Saint Peter's had a remarkable influence on later generations of designers, especially during the baroque period. This plaza (with its later modifications) was a great example of the use of visual reference points in organizing large urban spaces. As Bacon (1974) has suggested, the use of obelisks and a multinodal street design network (similar to the design of Washington, D.C., which was influenced by the baroque style) created a more complex urban geometry than that achieved by the simple block/gridiron pattern of the classical and Renaissance periods. The new geometric urban network embodied both order and centralization, while promoting movement and possibilities for spatial growth. Monumental architecture was no longer the terminus of major streets but the starting point, from which the city radiated in multiple directions.

Beyond design principles, three factors shaped baroque cities. First, port cities continued to act as major centers of growth—a pattern that had already emerged during the late medieval period. From northern Germany to Holland, England, Italy, and, even more strikingly, the United States, port cities flourished between 1500 and 1800 (Vance 1990). The post-Renaissance mercantile cities were of necessity ports, since international transportation was limited to oceanic voyages. The age of exploration, which began in the late fifteenth century, brought economic stimulus to cities across Europe. In fact, not only cities such as Antwerp, Amsterdam, and Bristol but also port cities in the eastern United States, such as Boston, benefited from the expanded global market and transportation network.

The new mercantilism reorganized the ports into layers of warehouses and labor residences and encouraged the gradual separation of social classes. This gradual sociospatial differentiation was made possible by the second factor that shaped baroque cities—land rent values. Vance identifies three distinct rent scales for commercial, working, and leisure-class needs. While commercial interests would need areas around ports that were also accessible to customers (internal and external), working people would be mainly concerned with living in places close to their work, and the leisure class would focus on specific areas (even streets) that had certain prestige. It is through this differential land rent value assignment that baroque cities began to display their third recognizable morphological factor—emerging segregation patterns. By the late seventeenth century, London and Boston were among many other European and American cities clearly witnessing spatial inequality.

One form of spatial polarization appeared with the burgeoning of the elites in the countryside. Social status associated with leaving the city carried a subtext of retreat to nature, as well.[9] During the baroque period the city was spatially reimagined to incorporate nature, albeit in an ordered manner. European cities witnessed the birth of "art imitating nature" in this aesthetic epoch (Lesnikowski 1982). Green external spaces were carried into ornate interior spaces and furnishings. Princely cities such as Versailles and Karlsruhe ostentatiously displayed this ordered nature. In a "people's city," however, a return to nature, at least on this scale, was not possible. Even in cities like London, where the fire of 1666 would have made it possible for a greening of the urban space, the rebuilt city became more mercantile, with much more attention to economic needs and rational land-use planning than to creating green space. While the West End of London could be designed to accommodate the well-to-do, the East End had to accommodate the masses and reflect the geography of their everyday existence. For the merchants and the labor classes, overall, the city remained a working place with its sites of production, warehousing, and differentiated residential areas.

If Versailles and other icons of the baroque period trigger images of class conflict, we must remember that the underlying philosophy of urban design during this period was rationalism, rooted in the optimism

---

[9]Toward the end of the Roman Empire, similar patterns of building villas in aesthetically pleasing settings, away from Rome, also occurred. However, the idea of having a home in the countryside was largely abandoned until this period.

of the new scientific age of reason, discovery, and prosperity. Understanding cities as spaces of civil societies was the next step in the social evolution of urbanism and the creation of the urban ideal, and to this the city is indebted to the eighteenth-century Enlightenment. While Rousseau and other philosophers engaged in debates on the "social contract," architects such as Jacques François Blondel began to discuss the importance of social relevance in design. The rationalism of the Enlightenment appeared to be a departure from the formalism of neoclassical design in its embrace of the ordered motifs of the organic phase of urbanism. This resulted in the production of interesting experiments in spherical and other nontraditional architectural forms. In cities, the new design concepts effected little change in the sociopolitical structure of the city. Working-class discontent led to the French Revolution and to political instability in other European countries in the years to come.

INDUSTRIAL CITIES AND THE SEARCH FOR THE IDEAL URBAN FORM

The French and the American Revolutions constituted an important political development of the late eighteenth century. Both extended the language of the Enlightenment into politics, but, more important, they marked the emergence of the modern nation-state. In the late eighteenth century, industrialization, colonialism, idealism, and scientific rationalism combined to create an age of contradictions. Six millennia after the age of Uruk, cities were finally on the precipice of becoming the dominant force in human history. These twin revolutions also inaugurated an era of political discontent and a new way of looking at "nature." In the midst of sociopolitical upheavals and industrial transformation, Romantics lamented the loss of the imagined community of the countryside and its organic social, cultural, and physical forms.

By the early nineteenth century, industrialization was transforming European cities across western Europe. Cities created and altered through this process were profoundly different from their predecessors. The demand for labor created a gradual rural-to-urban migration that became even more pronounced as the century progressed. Between 1801 and 1844 the population of cities such as Birmingham and Sheffield grew from 73,000 and 46,000 to 200,000 and 110,000, respectively (Engels 1987). Leeds more than doubled its population between 1801 and 1831, from 53,000 to 125,000. A significant source of this urban growth was the rural population, and in this respect the Industrial Revolution became the source of a new urban revolution. However, the concentration of the growing semiskilled and unskilled labor in cities led to a drastic sociospatial restructuring of the city. Though the tradition of spatial segregation had begun with mercantilism, the new capitalist economy further separated the social classes. Transportation innovations only exacerbated this growing socioeconomic problem. In the mid-nineteenth century, Friedrich Engels wrote of London and other English cities:

> I know nothing more imposing than the view which the Thames offers during the ascent from the sea to London Bridge. The masses of buildings, the wharves on both sides, especially from Woolwich upwards, the countless ships along both shores, crowding ever closer and closer together, until, at last, only a narrow passage remains in the middle of the river, a passage through which hundreds of steamers shoot by one another; all this is so vast, so impressive, that a man cannot collect himself, but is lost in the marvel of England's greatness before he sets foot upon English soil.
>
> But the sacrifices which all this has cost become apparent later. After roaming the streets of the capital a day or two, making headway with difficulty through the human turmoil and the endless lines of vehicles, after visiting the slums of the metropolis, one realizes for the first time that these Londoners have been forced to sacrifice the best qualities of their human nature, to bring to pass all the marvels of civilisation which crowd their city; that a hundred powers which slumbered within them have remained inactive, have been suppressed in order that a few might be developed more fully and multiply through union with those of others. . . .
>
> Every great city has one or more slums, where the working-class is crowded together. True, poverty often dwells in hidden alleys close to the palaces of the rich; but, in general, a separate territory has been assigned to it, where, removed from the sight of the happier classes, it may struggle along as it can. These slums are pretty equally arranged in all the great towns of England, the worst houses in the worst quarters of the towns; usually one- or two-storied cottages in long rows, perhaps with cellars used as dwellings, almost always irregularly built. These houses of three or four rooms and a kitchen form, throughout England, some parts of London excepted, the general dwellings of the working class. The streets are generally unpaved, rough, dirty, filled with vegetable and animal refuse, without sewers or gutters, but supplied with foul, stagnant pools instead. Moreover, ventilation is impeded by the bad, confused method of building of the whole quarter, and since many human beings here live crowded into a small space, the atmosphere that prevails in these working-men's quarters may readily be imagined. Further, the streets serve as drying grounds in fine weather; lines are stretched across from house to house, and hung with wet clothing. (Engels 1987, 31–33)

Later, in the case of Manchester, when he describes similar geographies of poverty, Engels explains how the upper and middle bourgeoisie lived in regularly laid out streets outside the impoverished areas of the center, while the richest of the rich lived in remote villas surrounded by sumptuous gardens. His observation of this segregation, and of the concealment of poverty, is still chilling today:

> And the finest part of the arrangement is this, that the members of this money aristocracy can take the shortest road through the middle of all the labouring districts to their places of business without ever seeing that they are in the midst of the grimy misery that lurks to the right and the left. For the thoroughfares leading from the Exchange in all directions out of the city are lined, on both sides, with an almost unbroken series of shops, and are so kept in the hands of the middle and lower bourgeoisie, which, out of self-interest, cares for a decent and cleanly external appearance and can care for it. True, these shops bear some relation to the districts which lie behind them, and are more elegant in the commercial and residential quarters than when they hide grimy working-men's dwellings; but they suffice to conceal from the eyes of the wealthy men and women of strong stomachs and weak nerves the misery and grime which form the complement of their wealth. (Engels 1987, 46)

The situation in the United States was no better. David Ward (1989) has documented the case of urban poverty and the impact of capitalism on American cities of the nineteenth century. It appears that the dynamics of capitalism, combined with the logic of efficiency, mobility, and industrial land-use needs, quickly translated into highly segregated cities that compartmentalized urban space according to function as well as by race, class, and ethnicity. While the Romantics found the industrial era an abomination and mourned the loss of "nature" and "community," economic forces affected the urban form, and the money-ordered city was born. From the city of gods to the city of people, the birth of the capitalist city signaled the demise of *Gemeinschaft* and all that it meant to intellectuals of the late eighteenth century and beyond.

In discussing this antiurban intellectual tradition in the West, Morton and Lucia White suggested that the dismay expressed by the Romantics over what the industrial city had done to rural life, nature, wilderness, and humanity was transformed years later into disappointment with the failure of cities to live up to the promise of civility (White and White 1962). The progressive agenda of the late nineteenth and early twentieth centuries grew out of the hope of recapturing those possibilities through planning and policy.

The industrial city was indeed both a beacon of hope for the masses, who flocked to its expanding employment and economic possibilities, and an enigma to better-off residents who frowned on the urban pathologies it produced. From New York to London, there appeared little chance of ameliorating the poverty and health problems in the growing inner-city slums of the nineteenth and twentieth centuries. Despite these social problems and the lack of adequate housing, rural people continued to move to cities. Additionally, American cities never lost their attraction for the immigrant population. This was partially due to the promise that the city held for the rural population. Despite the nostalgic assumptions of intellectuals, life in rural areas of Europe, and for that matter of the United States, was full of hardship and struggle. No wonder, then, that this period saw increased migration from country to city.

Industrial cities, however, were more than overgrown Renaissance or medieval cities. These cities benefited significantly from and were affected by innovations in construction as well as in methods of manufacturing, transportation, and planning (Whitehand 1977). Construction innovations profoundly affected how cities grew horizontally and vertically. Before the end of the nineteenth century, the use of structural steel made it possible to pack more people into smaller amounts of space. The invention of the elevator and the use of reinforced concrete contributed further to this process. Equally important for the growth of cities were improvements to transportation routes and the rapidity with which roads and streets were built. Other transportation innovations helped the expanded road network make the suburbanization of the middle class and the horizontal growth of cities a serious possibility. Initially the rail network, but eventually also trucks and automobiles, created the modern city that we recognize today.

Though Vitruvius and Greek planners before him began their urban plan with the careful layout of streets, the birth of modern cities in the nineteenth century increased the importance of streets. This trend is exemplified by Otto Wagner, one of the founding fathers of modern urban planning, who used transportation networks as an organizing tool for creating the modern city.[10] Wagner celebrated technology and looked with dismay on backward-looking attempts to solve modern urban problems. A pragmatist, he viewed the growth of cities as inevitable and therefore saw it as the ethical obligation of the designer to consider the immediate needs of the growing workforce and other

[10]See Schorske (1981) for a detailed discussion of urban modernism.

urbanites. This made the arrival at the modular design solution a reasonable expectation. In fact, Wagner's urban design solutions were modular from the scale of buildings to that of the city. He considered subdividing Vienna into semiautonomous districts of 100,000 to 150,000 people—each with its workplaces, modular apartments, and ordered public spaces. This was a utilitarian view of the modern city, but one that, whenever possible, also put due emphasis on creating pleasing aesthetics. The Wagnerian urban space was a celebration of modernity without a retreat to historicist monumentalism, which celebrated the culture of social elites and their heroes.

The irony of the nineteenth century, however, was the perpetual concern with what was being lost—nature, community, and human dignity—in the face of capitalist urban reality. If Wagner was unabashedly creating a city for the world of business, others, such as Camillo Sitte, were attempting to retain the organic quality of imagined medieval communities, with a deep commitment to history. The nation building of the nineteenth century required the schizophrenic urban design attitude of a belief in progress while gazing backward, a schizophrenia paralleled by the creation of more factory jobs while worrying about the loss of handicrafts. Labor issues alone pitted labor agents and labor advocates against each other. The growing pains of the modern age shaped the urban morphology of Western cities. Yet it remained (and still remains) unclear what the ideal city was to be. The industrial city concept produced the likes of Manchester and Chicago, massive industrial complexes that attracted millions of workers from near and far. While cities grew as fast as innovations in transportation and construction allowed, the question of how to manage the sprawling metropolis remained a persistent challenge.

By the time the Chicago School proposed its concentric zone model, based on Chicago of the 1920s, the city had become truly layered into social classes whose distance from urban externalities was determined by class, race, and ethnicity. The urban complex was shaped as much by sociocultural factors as by the economy and transportation network. By World War II, when the West had become fully urbanized, Hoyt's sectoral model of American cities illustrated how spatial segregation had become a part of the urban landscape and how the transportation network would make these patterns permanent. Faster commuting possibilities and innovation in housing technology had made the rapid expansion of the American suburbs a reality.

The suburbanization of the postwar period in the mid-twentieth century, though portrayed as the creation of bedroom communities, was largely the antecedent of the fragmented city—the multiple-nuclei model of Harris and Ullman, as well as the edge city of Garreau. The forces of the capitalist economy had a profound impact on the birth of the advanced capitalist city, with its decentered urban form. Ironically, from the end of the nineteenth century until the beginning of the twenty-first, design principles such as the City Beautiful Movement or the garden city have attempted to create an alternative urban ideal rooted in aesthetics, environmentalism, and moral order. However, following Wagner's assessment of these planning approaches as unrealistic and a form of escapism, the massive urban growth that occurred after World War II has remained focused on economic needs, including efficiency. It appears that, as in the new urbanism, green spaces and environmental considerations may carry a price tag that will lead to further class distinctions between those who can afford the recurring themes of "environmental ethics" and those who cannot. What remains a reality for the modern city are its diverse problems of traffic, air pollution, inadequate water, poor urban services, and overall failing health. We cannot abandon ideas of sustainability, equity, and livable spaces in our cities. Hopefully, these principles will form the basis for the next morphological transformation, shaping the cities of the twenty-first century.

FROM UR TO POSTMODERNISM: URBAN FORM(S) IN PERSPECTIVE

Thousands of years after the third dynasty of Ur, we can hardly find in Chicago, New York, Amsterdam, or Paris the urban form that dominated the Mesopotamian region. For that matter, it is difficult to find any semblance of Athens in them (short of some architectural mimicry of columns and graphic icons). But there appear to be some common morphogenic threads in the Western urban form—mobility, organization, and purpose. While environmental aesthetics and concerns for the destruction of the environment began with the Romantics (albeit informed by the likes of Epicurus and Spinoza) in the late eighteenth and early nineteenth centuries, urban parks, green beltways, and the greening of the city became common only after the mid-nineteenth century. Ironically, as technological innovations reduced the significance of environmental factors in determining the location, size, and growth patterns of our cities (apparent, for example, in the growth of cities in extreme climates), concerns for the destruction of the environment gradually grew into a forceful movement in the late twentieth century. The convergence of social and economic inequities with urban environmental problems has also brought us to the current concern with environmental justice. With global attention to population growth, urbanization, and the haphazard treatment of natural resources, we may be on the cusp of making sustainability the cornerstone of the twenty-first century's contribution to urbanism. A factor in the creation of the postmodern urban condition is the realization that the modern age brought us a culture of consumption that relies on scientific rationalism to promote progress at the expense of the environment and has contributed to an epistemological and ontological duality of city and environment. Disconnected from history and unclear about the meaning of nature and the environment, we remain unclear about what direction the city will take in the future. Will we move toward embracing environmental ethics and human rights, or will our excessive emphasis on localism and nationalism in economic development deter us from solving our common problems? At this fork in the road of cultural evolution, we can be assured of only one thing. The future of the world is urban, and the problems that cities and their environments face must therefore be addressed.

Since mobility, purpose, and organization remain the forces that shape our cities, we may wish to reconsider these factors in our planning and design efforts. Cities are more than collections of buildings. Movement provides the connection as well as the aesthetic quality and economic functionality that are needed to make the city work. From the Panathenaic Way to the Champs-Elysées to Broadway, the direction and purposive nature of movement define the urban experience. On smaller scales, each neighborhood is defined by its own set of streets and alleyways. Whether filled with cars or horse buggies, the personal and economic vitality of each city is defined by its road network. Romans under Julius and Augustus Caesar understood this when they paid close attention to traffic laws, as did other cities throughout the globe. We are still faced with the same set of transportation issues, albeit at much higher spatial scales, in most cities. Connectivity is the vital element of our social, economic, and political survival. Therefore, not unlike Haussmann, we remain concerned with the street—the transportation network—and what it may mean to urban governance. Combining land-use plans with transportation as an organizing principle, we attempt to create the purposive urban ideal that we hope will provide our desired quality of life. The degree to which we remain haunted by inequality is embedded in the nature of our social relationships, which create one of the most powerful morphogenic elements of our cities. Urban morphology remains a powerful narrative against which the urban ideal can be checked. To that end, urbanism and its changing spatial structure remain a part of our discourse on who we are, who we wish to be, and how the images of the environment may be embedded in our daily lives.

2   Population, Urbanization, and Environment

POPULATION GROWTH

IT IS DIFFICULT to discuss population without referring to three interrelated dimensions: magnitude, space, and time. Anyone engaged in population policy matters knows that one cannot make a convincing argument about population growth without some reference to the spatial and temporal specificity of growth, which highlights the differential nature of population impacts from one place to another. Beyond this methodological specificity, however, population discussions must also focus on the sociopolitical, cultural, historical, and economic contexts within which specific demographic patterns occur. From John Graunt's "life table" (1662)[1] of London to the most recent publications of the United Nations, demographic publications have touched upon the complexity of population through this multidimensional approach. Discussions of population "catastrophes" must likewise take into account time, space, politics, economy, and culture. One need only read Malthus's nineteenth-century treatise on population to appreciate the degree to which the sociopolitical and cultural values of a community determine its perception of population issues and of how they can best be managed. The looming population problems that confront us at various geographic scales are complex, and finding solutions is no easy task.

Given that our focus here is on cities, population issues will be examined within this context. Not only have cities acted throughout history as geographic nodes around which people have gathered in large numbers, but in the last few centuries they have also gradually become the major areas of human occupation. Urbanization rates in a number of countries exceeded 50 percent in the twentieth century, suggesting that people have looked to cities as places of work and leisure at a faster pace than ever before in human history. Cities have thus become arenas in which the dynamics of population are played out.

[1]For a discussion of Graunt's importance to demographic studies, see Szreter (2001).

POPULATION AND URBAN HISTORY

Historically speaking, the largest concentrations of human population have occurred in cities, and it is in the course of managing these higher population densities—from resource delivery to management—that city-states and what we call civilization have arisen. Among the earliest cities, the Sumerian city of Uruk had a population density of 100–150 persons per hectare as early as 3300 B.C. (see Table 2.1). This is similar to the population density of Paris in the early nineteenth century. For a city of such magnitude, a functioning infrastructure was a necessity. It is in understanding the demography of cities, in ancient and modern times alike, that we can begin to discuss how such places function and how their affairs are planned and implemented. Uruk may inspire images of Gilgamesh, but its twenty to thirty thousand inhabitants needed water, food, markets, garbage collection and clean up, and many other daily goods and services that are usually absent from our images of the past.

An important aspect of early cities was the physical limitation imposed upon them by the construction of walls. Growth beyond a certain level was unmanageable and therefore prohibited. The practice of prohibiting large numbers of visitors and merchants from staying for long periods of time in a city was common even in medieval Europe. As Table 2.1 illustrates, a city's population growth was closely tied to its size, as determined by the location of its walls. In the case of Uruk, that meant a proportional growth of three hundred hectares and twenty

TABLE 2.1. Estimated Population of Various Historical Cities

Decade	City Name	Size (in hectares)	Estimated Population
3300 B.C.	Uruk	200	20,000–30,000
2800 B.C.	Uruk	550	40,000–82,500
2100 B.C.	Rakhigarhi	80	8,000–12,000
	Harappa	150	15,000–22,500
	Mohenjo-Daro	250	20,000–37,500
	Ganweriwala	80	8,000–12,000
	Dholavira	100	10,000–15,000
1600 B.C.	Avaris	1000	100,000–150,000
1200 B.C.	Tyre	70	7,000–30,000
1360 B.C.	Chengchow	320	32,000–48,000
1360 B.C.	Erech	450	45,000–67,500

Source: Pasciuti and Chase-Dunn (2002).

to thirty thousand persons between 3200 and 2800 B.C. It was only after the demise of walls as a safety feature—due partially to the invention of gunpowder—that massive urban growth became physically possible.

The implicit connection between demography, physical structure, and urban services, including safety and protection, suggests the need for a managing body to assume the role of government. The emergence of political structure and economy is hence directly connected to population dynamics and the ability of humans to sustain (or increase) a reasonable population density over time. While the connection between population growth and urbanization is explicitly recognized, we must acknowledge that population growth is mostly a function of food production, medical innovation, and distributive infrastructures that determine its temporal and spatial patterns.

In this chapter we begin with a historical demography of the globe and its cities, though our focus is primarily on contemporary issues of population and how they affect cities. We offer a number of international case studies to highlight the differing nature of what we describe as "population momentum." It is important to keep in mind the differential impact of population in time and space. For example, one person in present-day Los Angeles consumes significantly more food and other goods than one person in India. We must therefore qualify the numerical aspect of population by considering energy consumption and natural resources. In this chapter we attempt to highlight these differences in order to illustrate the complexity of population issues. At the same time, we attempt to avoid the intellectual trap that many discussions of population fall into—namely, that population problems occur only in the developing world. While, numerically speaking, that is correct, from a resource consumption perspective it is not.

HISTORICAL DEMOGRAPHY

Inherent in the study of population is the assumption of data availability and accuracy, but, as students of antiquity are aware, historical demography is difficult to reconstruct and must involve both archival research and educated guesses. In most countries, including those in the West, systematic collection of population data did not occur until the early nineteenth century. In many ways, attention to the importance of population was inspired by the formation of modern nation-states, which began to reconceptualize city walls in the form of sharp boundaries for countries. These lines required formidable military forces for their protection and expansion. In the political environment of the late eighteenth century, when the Industrial Revolution and the formation of modern states led to a massive urban transformation of the world, Thomas Malthus's famous *Essay on the Principle of Population* (1798) occasioned significant debate on the importance of population to the future of the world, a question that was used in part to justify the creation of a brutal European imperialism in the nineteenth century and usher the world into the war-ravaged twentieth century. Malthus provided the world with something more than its first glimpse of future catastrophe caused by unchecked population growth. He also used what is known as natural theology to justify the outcome of population growth, which he saw as poverty and human misery. All inequalities were to be left alone to run their natural course. Malthus's theory of social evolution rested on a rudimentary concept of Darwinian natural selection in the context of political economy. His criticism of the English Poor Laws led to significant reforms in 1834, the year he died. Malthus contributed to an acute awareness of population problems, but his treatise met with serious criticism from a number of his contemporaries.[2] Engel's response appeared in 1844 in the *Deutsch-Französische Jahrbücher*, which was edited by Marx. The fathers of historical materialism argued that the notion of resource limitation (at least in the early nineteenth century) was ludicrous and that Malthusianism was nothing more than a bourgeois attempt to justify poverty and workhouses. As Foster (2000) argues, Marx's concept of the proletariat was a direct response to Malthus.

Regardless of its problematic content, discussed below, Malthus's theory of population created a sense of urgent need for population control, and a number of countries, including the United States, began to collect comprehensive population information. Though a basic census of the population was done as early as 1790, in 1820 the U.S. government began to gather information about its resident population, including immigrants, in greater detail. Outside the United States, with the exception of Scandinavian countries, which had begun their census taking in the early eighteenth century, England undertook its first census in 1801, but only in 1837 was the General Register Office (GRO) created to oversee the collection of demographic data in that country. The discipline of demography was a product of the nineteenth century. In fact, the word "demography" was first used in a Parisian publication in 1855 (Szreter 2001).

---

[2]For a detailed discussion of Malthus's theories and responses to them, see Foster (2000).

The birth of demography, the globalization of production and consumption, the creation and protection of nation-states, and concerns about available resources ensured that population issues would assume great importance.

POPULATION DYNAMICS: A GLOBAL HISTORY

Demographic studies have improved our understanding of how human communities grow numerically and structurally. We have striven to understand the environmental, political, cultural, and economic factors that alter and shape our communities, though much remains to be learned, as technological, medical, and environmental developments influence how we view reproduction and quality of life. Given the uneven distribution of such developments, politics and the uneven distribution of resources have a significant impact on population issues.

Before turning to the politics of population, however, it behooves us to look briefly at some basic population principles.

Population Dynamics

The Population Reference Bureau recently estimated that more than 106 billion people have been born since 50,000 B.C., when *Homo sapiens* probably first emerged (Haub 2002). Population growth was slow until roughly ten thousand years ago, when it began to mushroom with the Agricultural Revolution and the establishment of urban settlements. Table 2.2 provides a guesstimate of population magnitudes at various historical periods. Given the unreliability of estimates for past populations, however, this table serves simply as an approximation of the world's demography.

Table 2.2 clearly illustrates a number of population factors. First, even though the human population has grown significantly since the Industrial Revolution, it did not explode until the revolution in medicine and science of the twentieth century. Second, the decline in birth rates since the mid-eighteenth century correlates with higher rates of urbanization, population controls, and state-driven concerns about demographic issues. The significant decline in birth rates in the twentieth century is directly related to medical progress, economic stimulus, and population policies that have curbed human reproduction. Despite drastic decreases in birth rates, which also accompanied lower death rates in the second half of the twentieth century, the momentum created, mainly since the

TABLE 2.2. Historical Population Estimates

Year	Population (millions)[1]	Births per 1,000	Births between Benchmarks (billions)	Average Annual Births between Benchmarks (thousands)[1]
50,000 B.C.	—	—	—	—
8000 B.C.	5	80	1.1	27
A.D. 1	300	80	4.6	5,753
1200	450	60	26.6	22,159
1650	500	60	12.8	28,404
1750	795	50	3.2	31,719
1850	1,265	40	4.0	40,462
1900	1,656	40	2.9	58,005
1950	2,516	31–38	3.4	67,804
1995	5,760	31	5.4	120,607
2000	6,215	23	1.0	140,570
Number who have ever been born			106.5	—
World population in mid-2002			6.2	—
Percent of those ever born who were living in 2002			5.8	—

[1]Calculated by authors.
Source: Haub (2002).

end of World War II, has translated into what many see as a population time bomb. The last column in Table 2.2 vividly illustrates the concern with population growth. Whereas, at the dawn of the Industrial Revolution in 1750, the average annual growth rate approached 32 million, by 1950 it had more than doubled, to 68 million, and by the beginning of the twenty-first century, it had more than doubled again, to 141 million. At this rate, the global population will easily surpass 12 billion by 2050.

The historical pattern of population growth points to a conceptual and widely used model of "population transition." This model divides historical population growth within any one nation (but also arguably at the global level) into four stages (see Figure 2.1). During the first stage, higher birth and death rates result in low rates of natural increase, and population therefore grows minimally. This stage is characterized by a rudimentary agricultural economy, minimal urbanization rates, and primitive medical technology. The second stage is marked by a dramatic drop in death rates and stable or growing birth rates, which result in higher rates of natural increase. During this stage, modern technologies

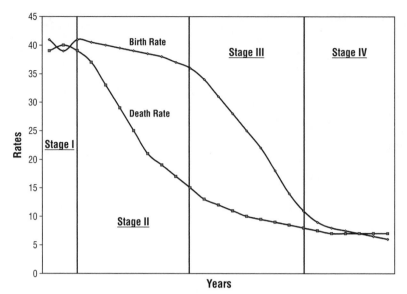

FIGURE 2.1. Demographic Transition Model.
Source: Figure by authors.

increase economic productivity, while medical services make possible a population explosion. The result is a relatively young population, whose difficulty in finding jobs in an increasingly urbanized country becomes a challenge to a heavily burdened political establishment. A significant number of developing countries are in this stage at present.

The third stage of demographic transition occurs when economic forces decrease the desire for larger families, resulting in dropping birth rates. Typically, this stage is accompanied by economic advancement, high rates of urbanization, and increasing numbers of women in the workforce. While the rate of population growth declines during this third stage, the fourth stage is characterized by minimal or no population growth. A number of countries within the former Eastern bloc, including Russia, Belarus, Bulgaria, Hungary, and Ukraine, had negative rates of natural increase in 2002 (Population Reference Bureau 2002). While the united Germany also achieved a negative rate of natural increase, other eastern European countries have had low birth rates for some time. In many of these countries, a combination of housing limitations, the necessity of two-income households, and the availability of legalized abortion as a family planning method has led to fluctuating, but generally lower, birth rates (Hornby and Jones 1993).

Apart from these few countries, most countries are growing rapidly, albeit with some declining rates of natural increase. In 2002 world population was growing at a rate of 1.3 percent annually (Population Reference Bureau 2002). At this rate we can expect a world population that exceeds 9 billion by 2050. Given the current dynamics of population geography, a significant majority of this growth will occur in less developed countries. Nigeria is expected to be the sixth-largest country in the world, in terms of population, by 2050. Other African nations will be in the top ten by 2050; the Democratic Republic of Congo could easily become the ninth-most populated country, and Ethiopia could be the tenth. The most populous country in the world in 2050 is likely to be India, which will probably reach 1.6 billion, surpassing China's projected 1.4 billion. Pakistan, Indonesia, Bangladesh, the Philippines, and Vietnam are projected to be the other large countries in Asia, also exceeding the 100 million mark. While Japan is also expected to achieve a population of 100 million, its population will probably decline between 2025 and 2050. In contrast, Iran's population at midcentury of slightly less than 100 million will still be growing.

POPULATION, URBANIZATION, AND CONSUMPTION OF RESOURCES

In 2003, as the world approached a 50 percent urbanization rate, its population growth rate was troubling, in that the hidden dimension of this demographic transition is a growth in energy consumption. Urbanization not only creates points of mass consumption; within cities, per capita energy consumption also escalates. This means that as various countries shift from the second to the third stage of demographic transition, their population growth slows down while their energy consumption increases. This is directly related to urbanization and economic growth, the very factors many assume reduce the rate of natural increase. In fact, while the correlation between gross domestic product (GDP) and urbanization rate has been documented (see, for example, Moomaw and Shatter 1996; Hornby and Jones 1993), the correlation between population growth, urbanization, and energy consumption needs to be further explored.

Two recent publications (Burney 1995 and Reddy and Balachandra 2003) point to the connection between GDP and urbanization rates and, to some degree, energy consumption. In order to explore the explicit connection between these variables, we combined population data from the Population Reference Bureau and the World Bank for 1998, the last year for which we had data.

We began our analysis by running a correlation between gross national income per capita (GNI) and urbanization rates, using the latest data for 2002. We chose to use this indicator, as opposed to GDP, because it was available from the same data source, and therefore a higher consistency in data could be expected. Furthermore, the Population Reference Bureau calculates GNI as the gross national income in purchasing power parity (PPP) divided by midyear population. GNIs are converted into international dollar rates, which indicate the amount of goods and services one can buy in the United States. This allows for better comparability of data across multiple countries.

As Figure 2.2 illustrates, urbanization is closely related to GNI (Pearson correlation of 0.69), but the impact of GNI on urbanization rate is higher during the early economic gains. Note that between 0 and $10,000, an increase in GNI translates to a significant increase in urbanization rates. At values higher than $10,000, however, gains in GNI have a more moderate influence on urbanization rates. This largely represents Western nations, whose urbanization rates reached high levels in the first half of the twentieth century. Notable exceptions are China and certain Middle Eastern countries, where GNIs and urbanization rates are not tied together as closely. Whereas in China high urbanization rates are achieved with minimal gains in GNI, in countries such as Kuwait and Qatar, increases in GNI rates cannot produce higher urbanization rates, since they already exceed 90 percent.

We can therefore conclude that while increases in GNI and urban growth are closely tied in stage two of demographic transition, further gains in urbanization will become more independent of increased wealth. That's when urbanization creates its own engine of production and consumption. In countries such as the United States, increases in the GNI can hardly translate to more people wanting to live in cities. Actually, as the GNI increases, the opposite could become true. One needs to look at expensive suburbs, golf courses, and resorts in order to understand how growth in income translates to consumption of "idealized" nature, nature, that is, that is organized and controlled.

An important aspect of the connection between GNI and urbanization rates is that a significant majority of countries fall within the lowest ranges of per capita income, while their urbanization rates increase. This indicates that, unlike Western countries, where increase in production led to higher GNIs and therefore potential for increased consumption, in a

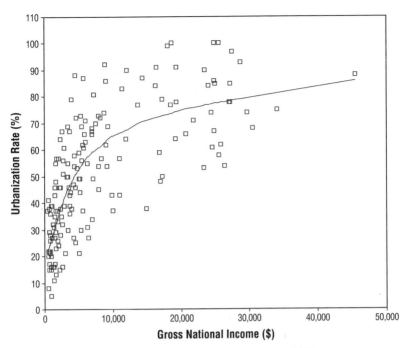

FIGURE 2.2. Gross National Income and Urbanization, 2002.
Source: Population Reference Bureau (2002). Figure by authors.

large number of developing countries urbanization rates outpace the
ability of the economy to generate jobs. Under these scenarios, cities in
developing countries turn into large centers of consumption, further
diminishing their ability to improve their economy and quality of life. In
order to illustrate the potential damage from such economically unjusti-
fied urban growth, let us turn to an analysis of energy consumption.

Using the World Bank data on commercial per capita energy consumption for 107 countries in 1997, we can assess the degree to which
energy consumption is related to urbanization rates and GNI. Figures
2.3 and 2.4 illustrate the results. Generally, it appears that the greatest
increase in energy consumption occurs when urbanization passes the
50 percent mark. This trend is not universal, however. In fact, a dual
trend appears at that point. Countries below the trend line include
Lebanon, Turkey, and Argentina. Uruguay is the most urbanized country with the lowest per capita commercial energy consumption. In
1997 China was still on the lower end of energy consumption compared to its urbanization rates, but the rapid expansion of industrial

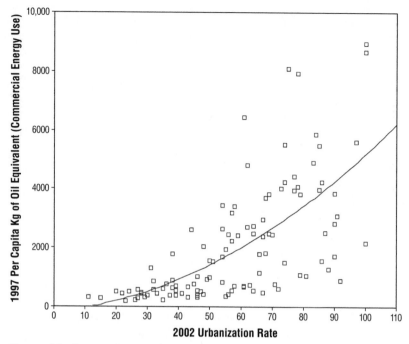

FIGURE 2.3. Energy Consumption and Urbanization.
Source: Population Reference Bureau (2002) and World Bank Data. Figure by authors.

and technological sectors in that country have resulted in further growth in its energy consumption.

Figure 2.4 illustrates a possible relationship between GNI and energy consumption. While not every country with a higher-than-average GNI is a large consumer of energy, this graph illustrates that, once again, the $10,000 mark seems to separate large and small consumers of energy. Close examination of the data suggests that a significant number of countries whose high urbanization rates did not lead to high energy consumption fall among those with lower GNI values. Hence we can conclude that a combination of urbanization and GNI rate has a more profound influence on the rate of energy consumption than either factor alone. A multiple regression analysis supports this hypothesis. With an R-squared of 0.71, energy consumption appears to be largely a function of the combined forces of urbanization and GNI. In other words, beyond settlement in urban places, a nation-state needs to have a wealth-generation engine that can vary from selling natural resources to production-based income.

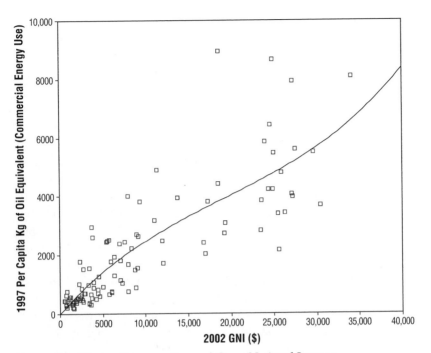

FIGURE 2.4. Energy Consumption and Gross National Income.
Source: Population Reference Bureau (2002) and World Bank Data. Figure by authors.

This will in turn increase the need to create a quality of life that becomes highly reliant on energy consumption. To this end, note the complexity of population growth. As countries become successively more urbanized and naturally aspire to improve their economic condition, they will increase their need for energy. This means that a person born in a nation with high urbanization and GNI rates will impose a higher energy cost on the globe than a person born in a less prosperous country. A person born into the twenty-first century will, on average, consume more energy than one born in the eleventh century. This means that while the policy tools of population control, namely, development and improved socioeconomic status, may result in a lower number of people, they do not guarantee lower total demand on world resources. While sustainability is now more and more widely seen as a solution to population issues, the twenty-first century, the age of global urbanization, may become a decisive moment in human history and how we handle global consumption patterns. Connecting the dots between urban and environmental concerns may become less a theoretical and moral debate and more an issue of human survival.

Between 1990 and 1997, while high-income countries increased their per capita commercial energy consumption by 1.1 percent, low- and middle-income countries in Asia, the Middle East, Latin America, and North Africa increased their consumption by rates ranging from 1.4 to 3.8 percent. During the same period, Europeans reduced their energy consumption by more than 5 percent. This is a reminder that prosperity and growth do not have to mean higher energy consumption. We do have a choice. However, global trends suggest that the initial phase of development, which coincides with stages two and three of the demographic transition, may be the most challenging period, when energy consumption and capitalist development could become the twin engines of social and environmental problems. From 1990 to 1997 the largest growth in per capita energy consumption occurred in upper-middle-income countries, which serves as a reminder that while stage three of demographic transition may bring a reduction in population growth, it by no means ensures better environmental prospects. From a public policy perspective, this suggests that population control programs need to be combined with conservation and environmental protection programs.

Although our analysis thus far has focused mainly on commercial energy consumption, it is also important to look at per capita energy consumption, or the energy used by individuals in everyday life. Petroleum used for operating motor vehicles and energy consumed to run household products such as refrigerators, cooling and heating devices, and home lighting, are equally important. Substantial research points to the rapid increase of total energy consumption at the individual level, especially as it relates to the growth of GDP and urbanization.

The United Nations Human Settlements Programme (UN-HABITAT) publishes detailed data showing "total energy requirement per capita." These data illustrate the inequitable nature of energy consumption. In 1994, for example, when Americans consumed 338 giga J per capita, Indians used 14 giga J per capita. In other words, on average, Americans used twenty-four times more energy than Indians did. Can this also be understood as meaning that as far as energy consumption is concerned, the American population, which totaled more than 260 million in the mid-1990s, is seven times larger than the Indian population of 930 million? Perhaps. The countries that occupy the moral high ground on population control, pointing fingers at the growing population of the developing world, should remember that in the mid-1990s 6.2 billion people could have lived in India with the same global energy consumption consequences that 260 million Americans produced.

In the mid-1990s, while the top energy consumers (with the exception of countries such as Kuwait) were highly developed countries, a significant number of Asian and African countries were among the lowest consumers of energy. Interestingly, the equation of development and energy consumption did not correlate perfectly, especially in the case of such countries as the United Kingdom and Switzerland, whose level of per capita energy consumption was significantly lower than that of the United States (their per capita energy consumption was 129 and 138 giga J, respectively). This suggests that energy consumption can be kept at a minimum without significant reduction in wealth or socioeconomic status.

In order to avoid higher energy consumption in Asian and African countries, sustainable development and energy conservation have to become important goals of public policy. To this end, the United Nations Environmental Programme (UNEP) has become more actively involved in formulating steps toward reaching such a goal. The Rio Summit, the Cairo International Conference, and the Kyoto Agreement represent attempts to establish common goals and guidelines for achieving a more sustainable future. As the Johannesburg Summit of 2002 showed, however, steps toward such a future are taken very slowly. Agenda 21 proved to be elusive, and many advanced and developing countries appear to have failed in their attempt to understand the balance between economic growth and environmental protection. The push for economic gain, especially in the developing world, may be in part due to a perspective that equates development (and even globalization) with social and economic equity (or democracy). In other words, the rush to technology and development become the tools for attaining democracy! This attitude was not necessarily born in less developed countries. While Lipset (1959) was among the earliest supporters of development as the means to democracy, staunch supporters of sustainable development have also advocated market-based approaches and progress in development and technology (for example, see Langeweg 1998). While development by itself does not necessarily reduce concern for the environment, the majority of developing countries face an expanding population, a need for raw materials, and an undying demand from their citizenry to increase the consumption opportunities equated with progress. That combination is detrimental to any attempt to protect the environment. A recent article by Ramanathan (2001) illustrates the problem by documenting that passenger-kilometers (PKM), or vehicle miles traveled (VMT) in the

United States, are growing faster than GDP and urbanization in India. In fact, it appears that the rise in PKM is closely tied to industrial growth. This suggests that while development positively affects the economic status of a country, the impact on the environment appears to outpace any gains. Furthermore, as Fincher (1998) argues, environmental degradation in the form of higher energy consumption and pollution is in many cases not the result of an increased standard of living, including energy consumption, for the local population, but of industrialization that benefits populations living elsewhere (e.g., the production of goods for export). Regardless of internal or external causes, the problems of increased energy consumption and environmental degradation are pervasive, and despite the lack of global collective will to correct these problems, individual countries such as India are pursuing initiatives of their own to correct them (see Reddy and Balachandra 2003; Chakrabarti 2001). This may be the kind of good news for which we should wait before rushing to embrace a utopian "green" future for the world. There are mileposts between the empirical and the normative world.

POPULATION AND HEALTH

Concerns about energy consumption and environmental degradation can be directly tied to health and urban conditions. Given the global importance of urbanization, issues of population become directly tied to this process. Despite the implicit assumption that a rise in urbanization is closely related to improved (or at least an increased desire for) access to health, healthy conditions are a rarity in a number of rapidly growing cities in the world. Nor is this situation confined to the developing world. Urban health occupies an important space in population discourse and policymaking the world over.

Analysis of mortality patterns in a number of large urban centers worldwide shows a less than promising picture of emerging health conditions. Surprisingly, in countries such as the United States and United Kingdom, the urban population has a shorter life expectancy than the rural (Geronimus 2001). In the late twentieth and early twenty-first centuries, this has mainly been due to the fact that inner-city regions are left entirely to minority populations, who are affected by a worsening of environmental conditions as well as by limited access to healthcare.[3]

---

[3]In 2000, most U.S. cities were populated primarily by minorities. In Los Angeles County, the non-Hispanic white population made up less than one-third of the 9.5 million people in this metropolitan area.

Socioeconomic and political inequity, combined with population dynamics that demand larger family sizes (Geronimus 2001), produce varying patterns of urban mortality in most American cities. This outcome is different from the earlier historical phase of urbanization, during which a shorter life expectancy was due to the spread of infectious diseases and unhealthy urban conditions. Whereas these early health problems could be remedied through medical and technological interventions, the current geography of inequities, which leads to unequal access to healthcare and healthy diets, demands solutions that run contrary to the logic of market economies in advanced capitalist states. Patterns of urban poverty, which have expanded in the past two decades (see Modarres 2002 and Jargowsky 2003), point to the worsening of urban conditions and the quality of life cities offer. Uneven population growth, urbanization, and health conditions are manifestations of deeper problems within a sociopolitical pattern. While it is disheartening to observe these growing urban problems, it is equally disturbing that, at the beginning of the new millennium, the only solution to these problems appears to lie in significant structural changes in society.

It is ironic that while the birth of cities was associated with a desire for safety, as well as improvements in economic, political, cultural, and social opportunities, cities have become symbols of fear, poverty, political corruption, and sociocultural decline. The reduction in environmental quality is the direct result of this decline, which has become the course of policy action in the past two decades in the United States and other Western nations. In the developing world, uncontrolled industrial growth, which treats cities as sites of inexpensive labor, has led to growing pollution, housing inadequacies, and other infrastructural shortcomings. The result has been a gradual decline in quality of life, which manifests itself in urban mortality rates and the emerging macro- and microgeographies of death.

A number of recent studies provide an interesting glimpse into global urban conditions. In January 2000 the Los Angeles County Department of Health Services and the UCLA Center for Health Policy Research published an important report entitled The Burden of Disease in Los Angeles County. This report used disability adjusted life years (or DALYs)[4] to

[4]DALY combines premature mortality, as measured by years of life lost, and morbidity, as measured by years lived with disability, to produce a single measure that allows for a more profound analysis of health conditions. This method was developed by the Harvard University School of Public Health with support from the World Health Organization and the World Bank.

reveal a number of findings regarding health conditions in the metropolitan area. First, it was clear that compared to using mortality rates alone, the new measurement pointed to chronic illnesses, drug and alcohol dependency, violence, motor vehicle crashes, HIV/AIDS, and suicide as major sources of health problems among the 9.6 million residents of Los Angeles County. Second, age-adjusted DALYs vary considerably by gender, race/ethnicity, and geography. For example, while homicide/violence was the second leading cause of DALYs in males, it ranked only twenty-fourth for females. Racially/ethnically, while African Americans and Native Americans had the highest age-adjusted DALYs, Latinos and Asian Americans had lower DALYs than whites. Numerically speaking, while coronary heart diseases cause the highest number of deaths and years of life lost (YLL), homicide/violence ranks second in terms of the second measurement. This is followed by trachea/bronchus/lung cancer, stroke, motor vehicle crashes, suicide, and HIV/AIDS. The leading causes of YLL are created mostly by urban conditions marked by high concentrations of population, poverty, poor dietary habits, and pollution. This becomes clear when one considers the differences in race/ethnicity and geography (see Tables 2.3 and 2.4). For example, while homicide/violence appears to create high DALYs among African Americans and Latinos, among whites and Asian Americans this phenomenon is entirely absent. In its place one finds a higher presence of diet-related and old age illnesses. Interestingly, while depression ranks ninth for whites and tenth for African Americans, among Latinos and Asians with a significant foreign-born population component, depression ranks third and second, respectively. Clearly, while urbanization and access to health care provide some indicators of health conditions, migration and overall well-being produce ramifications that go beyond simple medical treatment.

The geography of health, produced by various causal factors, including socioeconomic and nativity factors, leads to an interesting spatial pattern of DALYs in Los Angeles County. As Table 2.4 illustrates, while coronary heart disease produces the highest-ranking DALYs in six of the eight subregions of the county, in the south and the Antelope Valley homicide/violence and alcohol dependency create the highest DALYs, respectively. It is not surprising that the southern region is mainly occupied by a highly segregated, mostly low-income African American population and the Antelope Valley is home to a growing number of Latinos, military personnel, and, as the expanding urban fringe, middle- to lower-income populations in search of dwindling

TABLE 2.3. Leading Causes of Disability Adjusted Life Years (DALY) by Sex, Los Angeles County, California

	Male			Female		
		DALYs			DALYs	
Rank	Cause	Years	Rate	Cause	Years	Rate
1	Coronary Heart Disease	42,560	8.87	Coronary Heart Disease	30,326	6.27
2	Homicide/Violence	40,555	8.45	Alcohol Dependence	26,196	5.42
3	Alcohol Dependence	34,676	7.23	Diabetes Mellitus	23,598	4.88
4	Drug Overdose/ Other Intoxication	25,847	5.39	Depression	21,337	4.41
5	Depression	22,112	4.61	Osteoarthritis	19,828	4.10
6	Osteoarthritis	19,984	4.17	Stroke	17,388	3.60
7	Motor Vehicle Crashes	19,744	4.12	Alzheimer's/Other Dementia	16,987	3.51
8	Diabetes Mellitus	18,858	3.93	Breast Cancer	15,379	3.18
9	HIV/AIDS	17,596	3.67	Emphysema	14,843	3.07
10	Trachea/Bronchus/ Lung Cancer	16,646	3.47	Trachea/Bronchus/ Lung Cancer	13,139	2.72

Source: Los Angeles County Department of Health Services and UCLA Center for Health Policy Research (2000).

affordable housing. The geography of wealth, race, ethnicity, and nativity has clearly led to a geography of health that makes Los Angeles the poster child for what is to be expected in other urban settings. If Los Angeles is reproduced through a global urge for growth, the future of urban health can only worsen under market conditions that immediately translate to exaggerated geographies of wealth.

Given that China and India are the two largest countries in the world and collectively house one-third of its population, their mortality patterns are of special concern. Ru-Kang (1993) suggests that urban mortality rates in China have increased since 1981 owing to an aging population and worsening environmental pollution. Examining the case of Shanghai, China's large megacity, Takano et al. (2002) argue that while significant economic and population growth occurred in the 1990s, the city as a whole fared better than other cities, if age-adjusted mortality rates, rather than the crude death rate, are analyzed. The authors do point, however, to significant spatial variations of their standardized mortality rates at the ward or subregional level of the

TABLE 2.4. Leading Causes of Disability Adjusted Life Years (DALY) by Race/Ethnicity, Los Angeles County, California

Rank	Cause	DALYs Years	DALYs Rate	Cause	DALYs Years	DALYs Rate
	White			Black		
1	Coronary Heart Disease	44,010	13.63	Homicide/Violence	14,264	15.79
2	Emphysema	20,411	6.32	Coronary Heart Disease	11,380	12.60
3	Alcohol Dependence	19,933	6.18	Diabetes Mellitus	9,737	10.78
4	Trachea/Bronchus/ Lung Cancer	19,073	5.91	Alcohol Dependence	6,966	7.71
5	Alzheimer's/Other Dementia	18,533	5.74	Stroke	6,950	7.69
6	Diabetes Mellitus	15,287	4.74	Trachea/Bronchus/ Lung Cancer	5,526	6.12
7	Osteoarthritis	15,174	4.70	HIV/AIDS	5,514	6.10
8	Stroke	14,931	4.63	Asthma	4,712	5.22
9	Depression	14,369	4.45	Emphysema	4,388	4.86
10	Drug Overdose/ Other Intoxication	14,268	4.42	Depression	4,151	4.59
	Latino			Asian/Pacific Islander		
1	Alcohol Dependence	28,305	6.70	Alcohol Dependence	8,345	6.78
2	Homicide/Violence	23,927	5.67	Depression	6,151	5.00
3	Depression	18,530	4.39	Osteoarthritis	5,758	4.68
4	Diabetes Mellitus	15,116	3.58	Coronary Heart Disease	5,219	4.24
5	Osteoarthritis	14,864	3.52	Stroke	4,290	3.49
6	Motor Vehicle Crashes	14,222	3.37	Diabetes Mellitus	4,118	3.35
7	Coronary Heart Disease	12,207	2.89	Alzheimer's/Other Dementia	3,864	3.14
8	Drug Overdose/ Other Intoxication	9,542	2.26	Emphysema	3,015	2.45
9	Stroke	8,873	2.10	Motor Vehicle Crashes	2,745	2.23
10	Cirrhosis	7,994	1.89	Unintended Firearm Injury	2,487	2.02

Source: Los Angeles County Department of Health Services and UCLA Center for Health Policy Research (2000).

urban area. Illustrating the importance of environmental considerations such as green spaces, parks, and open spaces, they highlight the unequal distribution of such amenities across the metropolitan area. A city of 13 million, Shanghai manifests a social geography similar to other megacities worldwide, and so it is not surprising that mortality rates are much higher in some neighborhoods than in others and that such conditions correspond to socioeconomic status and environmental conditions.

While Shanghai is a megacity that gives significant attention to environmental conditions, cities such as New Delhi and Mumbai in India do not. Furthermore, the geography and magnitude of poverty in these Indian cities are significantly worse. In this these two cities appear to be better representatives than Shanghai of other cities in the developing world. According to data published by the registrar general and Census Commissionaire of India, while the crude death rate has declined in the past two decades, the urban death rate, though lower than the rural death rate, increased after 1993 and stabilized at rates above 6 percent. This figure, however, is a national average. In Delhi the crude death rate is significantly worse in poverty-stricken neighborhoods, where respiratory infections and diarrhea are responsible for more than one-third of deaths (Bhatnagar et al. 1988). In Mumbai, slums house more than half the population, more than 2 million residents have no sanitary facilities, and more than half suffer from chronic malnutrition.[5] Add air and water pollution, and the result is an urban environment in which high mortality, especially for young children of largely poor neighborhoods, points to an unbearable quality of life.

The situation in African cities is no better. Analyzing rural/urban mortality patterns for children under five in sub-Saharan Africa, with particular attention to Kenya, Gould (1998) suggests that the recent worsening of economic and environmental conditions has diminished the gains made by urbanization. In other words, rapidly growing larger cities in sub-Saharan Africa are plagued by a growing mortality rate that, unabated, might match rural conditions. Lalou and LeGrand (1997) find a similar situation in Mali, Burkina Faso, and Senegal. They argue that the absence of adequate healthcare services and deteriorating urban environmental conditions are increasing mortality risks and the prevalence of diarrhea, malaria, and respiratory infections. Gould refers to these

[5]See the Mega-Cities Project website on Mumbai, at http://www.megacitiesproject.org/network/mumbai.asp.

conditions, which disproportionately affect the poorer populations in rapidly growing cities of Africa (but also in Asia and other regions of the world) as the "new urban penalty."

Since 90 percent of future population growth is expected to occur in cities, and since the highest rates of urbanization are expected to appear in the developing countries of Asia and Africa (Livernash and Rodenburg 1998), the future of the urban environment under such structurally unbalanced and environmentally inadequate policy regimes appears deeply disturbing. However, as Nietzsche reminds us, what is apparent is not necessarily true. First, the poverty in a number of cities in the developing world is not generated independently of global processes. In other words, the consumption patterns of developed nations are partially responsible for the emerging patterns of inequity. This does not excuse the governments of developing nations from attempting to resolve their problems.

Second, patterns of inequity and the unequal impact of environmental problems on urban residents are not exclusive to cities of the developing world. Concerns with such environmental inequities have been growing, especially over the past two decades. Recent literature on environmental justice in the United States and the establishment of an office within the Environmental Protection Agency to respond to this concern reflect the changing discourse on the urban environment, even within advanced urban societies. While the topic of environmental justice will be addressed separately, here we simply note that while stationary sources of pollutants have historically affected major urban populations in the developed nations, within the past few decades the shrinking of the industrial sector and its flight to other regions of the world has reduced concerns about factory-produced pollutants in Western cities. Of course, in response to government pressure to reduce their pollution, some of the worst polluters have moved to job-hungry, less environmentally resistant countries. Furthermore, as the polluting industries of developed nations have moved elsewhere, the consumption patterns of these countries, which lead to higher energy demands and higher rates of mobility, have created new challenges.

A global perspective on environmental justice has become necessary, so that as American (or other nations') factories impose pollution and worsening environmental conditions on other countries to benefit consumers in the West, they should be held liable for any health ramifications to current and future generations in the affected countries. In addition, concern for the environment in developed countries, especially the United States, should shift from factories to mobile sources of pollution, mainly automobiles, and their uneven impact on urban residents.

There is evidence that this shift has already taken place. In the United States, the earliest concern with the impact of freeways on air pollution, especially in nearby areas, began in the 1960s. One of the first articles on this topic appeared in the journal Atmospheric Environment in 1975 (Whitby et al. 1975). Measuring aerosol samples along a major freeway in Los Angeles, the authors concluded that rush-hour traffic produced a significant adverse impact on the ambient air pollution along the targeted thoroughfares. This and other early papers on worsening of air pollution due to automobiles were only implicitly connected with urban growth and its negative impact on less well-to-do communities. Furthermore, they did not address the issue of lower land values and the concentration of rental property along major traffic arteries, which were increasingly becoming the hot spots of urban air pollution, especially of particulates. The only exceptions were found in discussions of land-use policies. For example, Smit (1984) suggested that the externalities, or adverse effects of automobile use on various neighborhoods, should be considered in formulating land-use policies.

Increased attention to environmental protection in the late 1980s and early 1990s led researchers to shift their focus to the connection between pollution and mortality rates (see, for example, Dockery et al. 1993; Moolgavkar et al. 1994; Vostal 1999). In addition, there was increased interest in the inequitable way in which pollutants affect minority communities, especially those of low socioeconomic status, and other susceptible demographic groups such as children and the elderly (Brown 1995; Wilson and Spengler 1996; Friedman et al. 2001). Although earlier research had made the connection between hazardous waste and stationary sources of pollutants and such communities (e.g., U.S. General Accounting Office 1983; Commission for Racial Justice, 1987), studies of mobile sources of pollution—mainly cars—have appeared more regularly since the early 1990s. Publications by Dockery et al. (1993) and Moolgavkar et al. (1994) are among some of the earliest studies of particulates and the geography of their health impact. While these studies were replicated in other Western countries (such as the U.K., Sweden, and the Netherlands—e.g., Nevalainen and Pekkanen 1998; Hoek et al. 2002), the case of the United States remains unique.

As reported by Donohoe (2003), for every gallon of fuel extracted, processed, and distributed, twenty-five pounds of carbon dioxide, along with other pollutants, are released into the atmosphere. In a country such as the United States, where there is one car for every two people, the energy consumed and the air pollution produced by cars represent a significant problem both locally and globally. With the rise of high-fuel-consuming cars like SUVs within the past decade, not only has American fuel consumption increased, but so has its contribution to air pollution. At an average of slightly more than twenty miles to the gallon, SUVs and light trucks consume significantly higher volumes of fuel than small sedans. According to the National Household Travel Survey, Americans' personal miles of travel reached 11 billion per day in 2001 (U.S. Department of Transportation 2003). Even with the average national standard of 27.5 miles per gallon for passenger cars, and assuming that everyone owns such an average automobile, the daily fuel consumption for the country would reach 400 million gallons. This translates to 10 billion pounds of carbon dioxide and other related pollutants per day. Given the unequal impact of air pollutants on various social classes within a city, it should come as no surprise that the environmental justice movement is targeting SUVs and trucks and will continue to do so in the future.

Furthermore, considering the high rates of fuel consumption and air pollution production by the American public, it is easy to see how, despite its relatively low rate of population growth, the United States produces a significantly larger proportion of automobile-generated air pollutants per capita globally, even with cleaner cars. This pattern of energy use and environmental degradation by one of the world's most developed countries leaves little hope that developing nations, who model themselves after the United States and its patterns of consumption, will be less harmful to the world's natural resources and its urban environment. If the highest rate of population growth is to occur in the developing world, especially its cities, it is crucial that we pay special attention to their energy needs and the negative consequences for the world ecology.

THE POLITICS OF POPULATION CONTROL AND SUSTAINABLE URBAN GROWTH

Even though cities—specifically urbanization—can be viewed as an important factor in reducing the global population growth rate, the fact remains that the majority of people who will be born in coming decades will be born in cities (Brockerhoff 2000). By 2025 the urban share of the global population will reach 58 percent. In that year, while the urbanization rate of more developed countries will have increased by less than 10 percent since 2000, the rate of urbanization in countries in developing regions of the world, mainly in Asia and sub-Saharan Africa, will have increased the most. With a growth of 14 percent in developing countries collectively, Asian and sub-Saharan urbanization rates will have increased from 35 to 50 and from 34 to 49 percent, respectively (Brockerhoff 2000). Put another way, most of the world's population growth between 2000 and 2025 will occur in the urban centers of the less developed countries, excluding Latin America, where urbanization rates are already high. By 2015 there will be more than four hundred cities in less developed countries with a population of 1 million or more, whereas in 1975 there were only 110 such cities in these countries. The expected number of large cities will be almost three times the number in more developed countries (Brockerhoff 2000). The United Nations predicts that the largest cities in 2015 will be, in order of size, Mumbai (India), Tokyo (Japan), Lagos (Nigeria), Dhaka (Bangladesh), Sao Paulo (Brazil), Karachi (Pakistan), Mexico City (Mexico), Delhi (India), New York (United States), and Jakarta (Indonesia), with a range of a high of 28 million (Mumbai) to a low of 17 million (Jakarta) (United Nations 2000). Note that seven of the top ten cities will be in Asia and Africa. This pattern of urbanization is expected to continue in the first half of the twenty-first century. By 2030 the urban population of Asia will be 2.6 billion, an increase of 1.1 billion since 2000. In Africa the urban population will triple in the first three decades of the century, from 297 million to 766 million (McGee 2001).

Given this population scenario, we can see an urban crisis on the near horizon. In fact, many of those concerned with such problems can simply point to Africa and Asia and tell us that the crisis has already begun. Whether through rampant poverty, food shortages, inadequate urban services, housing inadequacies, polluted water and air, or violence, rapid urbanization is already imposing a heavy toll on urban populations of these continents. The shift in focus from rural to urban health issues in sub-Saharan Africa by a number of health policy advocates reflects what Few et al. (2003) call the contextual contrasts between urban and rural health settings. The significant difference between the social and economic settings of rural and urban areas, especially in Asia and Africa, indicates a gradual disorganization of traditional societies through the urbanization process. In his Kafkaesque account of western Africa,

Kaplan (2000) warns of the looming problem of the loss of communal life in urban settings. With desertification and deforestation sending more people to urban areas, where the social organization of rural areas is lost to economic needs, cities become spaces of social dysfunction, anarchy, and violence. Recently, in the Congo, but also in some of the west African nations, such as Sierra Leone, violence is committed mostly by youth, who in the absence of social order and family ties have attained a level of lawlessness that can only be described as horrific. There is no denying that a postcolonial environment of unsustainable artificial boundaries and other colonial legacies have contributed substantially to this situation, but it would be a mistake to place exclusive blame on colonialism. Rapid population growth and continued environmental degradation, in the absence of economic growth and democratic sociopolitical systems, have taken a significant toll on various African and Asian countries. Kaplan reports on the dire conditions of two shantytowns of Abidjan in Ivory Coast, a country with a relatively better economy than its neighbors. If Chicago and Washington, the names of two poor neighborhoods of this city, show us the African urban future, it is easy to see sub-Saharan nations as the manifestation of Malthusian prophesies. But is this inevitable, an unpreventable outcome of rapid population growth, environmental degradation, overcrowding, and the spread of contagious diseases in growing urbanized regions of the developing world?

Given the consumer societies of the West and their unquenchable thirst for resources, it is difficult to envision a simplistic population control measure for the developing world alone. The problems of cities in developing countries are neither entirely external nor entirely internal. The answer may lie in a global goal of sustainable development that attempts to curtail energy consumption in the Northern Hemisphere and establish democracies, equitable development, and population control in the Southern. The United Nations' Agenda 21, the Kyoto Protocol, the Cairo Conference, and the Rio Summit were the first initial attempts to create a global vision and commitment to sustainable development. It is therefore crucial that we understand them, since they represent the current ideas of policymakers. More important, the failure of such initiatives, especially in such countries as the United States, should be studied, for they illustrate the insularity of some of the most developed nations in the age of globalization. Furthermore, we must keep in mind that population controls are not simply a means to an end. Despite the global slowing of population growth, many of the external forces that exacerbate population growth remain strong. Therefore, the goals of sustainability should move beyond a focus on the decline in crude natural rates of increase. Let us consider the unique case of India, a country with one of the earliest family planning programs.

WHAT HAPPENED IN INDIA?

In a few decades, India will have the distinction of being home to two of the largest cities in the world, when Mumbai and New Delhi surpass the 20 million mark. A country troubled by a dysfunctional rural economy and declining urban conditions must expect a crisis in the urban environment. With population expected to reach 1.4 billion by 2025 and 1.6 billion by 2050 (Population Reference Bureau 2002), Indian cities are becoming places where millions of people will be looking for scarce jobs, and while they wait they will expect to receive urban services such as healthcare and education. Air pollution from industries and motor vehicles is already creating adverse environmental conditions in many large Indian cities, and adequate access to safe drinking water in cities like Madras is already a challenge (Visaria and Visaria 1995). As in other developing countries, despite the enactment of a number of environmental standards, policymakers and urban administrators in India find it difficult to balance job needs and development with environmental conservation and protection. Cities continue to draw rural populations in search of work, regardless of their worsening environmental conditions. During the first decade of this century, the urban population of India could easily surpass 300 million, which is larger than the entire current population of the United States. India's growing population, despite a five-decade-old family planning program, is rolling this country down a steep hill into economic and environmental crisis. As of early in the 1990s India had achieved only small gains in controlling its population growth. By 2006 the rural population explosion is transposing itself to the cities. Despite the famous claim of the Indian delegation to the 1974 Population Conference in Bucharest that "development is the best contraceptive," gains in technology and development did not necessarily result in a rapid decline in the birth rate in India—or for that matter in many developing countries. One can even argue that gains in population control are more than made up for by increases in consumption.

An important lesson to be learned from Indian population issues and urbanization patterns is their extreme unevenness in population distribution and development. More than 40 percent of the population lives in the northern states, which extend from Rajasthan to West Bengal. These states, including Uttar Pradesh, home to 166 million people

(Registrar General and Census Commissioner, India 2001), represent some of the fastest-growing areas of India. Uttar Pradesh and neighboring Bihar are also heavily cultivated. From 1991 to 2001—the time span between the two Indian censuses—Uttar Pradesh and Bihar grew 26 and 28 percent, respectively. The northern states, including these two, have a larger percentage of poor inhabitants, as well as a lower life expectancy. Many of their cities, including Delhi, Calcutta (province of West Bengal), and Ahmadabad (Gujarat) have large slum sectors that house more than one-third of each city's population. This phenomenon is not exclusive to large cities—places like Cuttack, in the province of Orissa, have as many as forty-nine major slum pockets housing a significant portion of the urban population (Routray and Pradhan 1989).

During the past decade, the provinces of Kerala (9 percent), Tamil Nadu (11 percent), Andhra Pradesh (14 percent), and Karnataka (17 percent), in southern India, had the lowest rates of growth. The answer to why these states have a slower population growth rate is not a simple one. While Kerala has a literacy rate of 91 percent, Tamil Nadu, Karnataka, and Andhra Pradesh have achieved 73, 67, and 61 percent literacy rates, respectively (Registrar General and Census Commissioner, India 2001). The answer lies beyond simple measurements of progress. What distinguishes Kerala and Tamil Nadu from other states is a combination of higher literacy rates, especially female literacy rates, a high female-to-male ratio, and lower rates of urbanization (see Table 2.5). Kerala not only has an unusually high literacy rate, but its population has also urbanized at a slower pace, which may suggest a less disruptive growth pattern. Though Tamil Nadu houses Madras, a relatively large urban conglomeration, Kerala has few major cities.

As Kaplan (1996) has observed, India appears to run a course contrary to Malthusian prophecy. Having experienced a significant gain in its per capita GNP since the early 1980s, the country has witnessed extensive technological development and impressive foreign investment, and exports food to other countries. Yet in Bihar and other less well-to-do provinces, population control has become an enigma, female literacy rates are abysmal, runaway industrialization has meant serious environmental problems,[6] and cities appear to be fatigued from traffic,

---

[6]As a result of rapid industrialization and lax attention to the proximity of population to hazardous chemical sites, a number of potentially dangerous situations have been created. While Bohpal has become synonymous with industrial disasters, other cities in India (and in other countries) suffer from a quiet poisoning of their population. For a case study on Hyderabad in Anhra Pradesh, see Sekhar et al. (2003).

pollution, and a diminishing quality of life. In a country exploding in population, jobs are far more important to the political and economic well-being of the nation than concerns about such urban externalities. The logic of "taking care of the present" leaves little consideration for tomorrow. India's investment has at best attempted to respond to its population geography, which has meant uneven development. Many Indians have to commute long distances to job centers,[7] creating extreme transportation problems despite an extensive, operational, and fully utilized railway. What appears to be lacking in India is consideration for sustainability in the bureaucratic planning process. This does not mean, however, that researchers and ecological activists have not attempted to write about or implement unique programs in various areas of India.

Kaplan provides a glimpse into an environmentally friendly program in Rishi Valley in Andhra Pradesh (Kaplan 1996), a province with relatively low population growth but home to Hyderabad, a major urban complex with a population of 5.5 million. Rishi Valley is largely the legacy of a twentieth-century Indian naturalist philosopher, Jiddu Krishnamurti. Even though Kaplan refers to his ideas as "green culture," "earth nationalism," and a precursor to the "Gaia" theory, Krishnamurti was a practical philosopher who built an elite boarding school in a secluded valley in early 1926. Rishi Valley is located about 135 kilometers northeast of Banglore, a major city in Karnataka province. The school is unique in that it was built not in a picturesque setting but in a barren landscape, where students could engage in healing the land with villagers. This was an attempt to bring the city to the villagers and introduce modernity to rural settings. Kaplan quotes a teacher at the boarding school, who explains that this philosophy was embedded in a belief that a country cannot be modern if its villages are medieval. Of course, by "modernity" the school did not mean industry; its goal was to heal the land through careful planting. According to Kaplan, schoolchildren plant twenty thousand trees and shrubs each year and distribute one hundred thousand seedlings in the valley. Using various environmental techniques and the manipulation of topography and soil, students engage in recharging the water table in the valley and diminishing the need for fertilizers. As one of the teachers explained to Kaplan, education in Rishi Valley is a deep learning process that

[7]For an interesting case study of how rapid industrial development in Hosur (in Dharmapuri) has led to a large commuting workforce, see Hornby and Jones (1993).

TABLE 2.5. Census of India, 2001

India/State/	Total Population			Urban		
Union Territory*	Total	Males	Females	Total	Males	Females
INDIA	1,027,015,247	531,277,078	495,738,169	285,354,954	150,135,894	135,219,060
Andaman and Nicobar Islands*	356,265	192,985	163,280	116,407	64,148	52,259
Andhra Pradesh	75,727,541	38,286,811	37,440,730	20,503,597	10,434,632	10,068,965
Arunachal Pradesh	1,091,117	573,951	517,166	222,688	120,391	102,297
Assam	26,638,407	13,787,799	12,850,608	3,389,413	1,804,642	1,584,771
Bihar	82,878,796	43,153,964	39,724,832	8,679,200	4,643,278	4,035,922
Chandigarh*	900,914	508,224	392,690	808,796	451,387	357,409
Chhatisgarh	20,795,956	10,452,426	10,343,530	4,175,329	2,161,443	2,013,886
Dadra and Nagar Haveli*	220,451	121,731	98,720	50,456	29,844	20,612
Daman and Diu*	158,059	92,478	65,581	57,319	28,902	28,417
Delhi*	13,782,976	7,570,890	6,212,086	12,819,761	7,037,671	5,782,090
Goa	1,343,998	685,617	658,381	668,869	345,991	322,878
Gujarat	50,596,992	26,344,053	24,252,939	18,899,377	10,054,630	8,844,747
Haryana	21,082,989	11,327,658	9,755,331	6,114,139	3,310,036	2,804,103
Himachal Pradesh	6,077,248	3,085,256	2,991,992	594,881	331,005	263,876
Jammu and Kashmir	10,069,917	5,300,574	4,769,343	2,505,309	1,374,728	1,130,581
Jharkhand	26,909,428	13,861,277	13,048,151	5,986,697	3,200,847	2,785,850
Karnataka	52,733,958	26,856,343	25,877,615	17,919,858	9,237,750	8,682,108
Kerala	31,838,619	15,468,664	16,369,955	8,267,135	4,017,879	4,249,256
Lakshadweep*	60,595	31,118	29,477	26,948	13,922	13,026
Madhya Pradesh	60,385,118	31,456,873	28,928,245	16,102,590	8,481,617	7,620,973
Maharashtra	96,752,247	50,334,270	46,417,977	41,019,734	21,891,032	19,128,702
Manipur	2,388,634	1,207,338	1,181,296	570,410	283,910	286,500
Meghalaya	2,306,069	1,167,840	1,138,229	452,612	228,037	224,575
Mizoram	891,058	459,783	431,275	441,040	226,065	214,975
Nagaland	1,988,636	1,041,686	946,950	352,821	195,035	157,786
Orissa	36,706,920	18,612,340	18,094,580	5,496,318	2,900,487	2,595,831
Pondicherry*	973,829	486,705	487,124	648,233	323,119	325,114
Punjab	24,289,296	12,963,362	11,325,934	8,245,566	4,462,715	3,782,851
Rajasthan	56,473,122	29,381,657	27,091,465	13,205,444	6,987,178	6,218,266
Sikkim	540,493	288,217	252,276	60,005	32,831	27,174
Tamil Nadu	62,110,839	31,268,654	30,842,185	27,241,553	13,759,669	13,481,884
Tripura	3,191,168	1,636,138	1,555,030	543,094	276,850	266,244
Uttar Pradesh	166,052,859	87,466,301	78,586,558	34,512,629	18,369,536	16,143,093
Uttaranchal	8,479,562	4,316,401	4,163,161	2,170,245	1,173,021	997,224
West Bengal	80,221,171	41,487,694	38,733,477	22,486,481	11,881,666	10,604,815

(continued)

TABLE 2.5. *(continued)*

India/State/ Union Territory*	Rural			Literates		
	Total	Males	Females	Total	Males	Females
INDIA	741,660,293	381,141,184	360,519,109	566,703,280	339,905,576	2226,797,704
Andaman and Nicobar Islands*	239,858	128,837	111,021	252,945	146,536	106,409
Andhra Pradesh	55,223,944	27,852,179	27,371,765	40,364,765	23,636,077	16,728,688
Arunachal Pradesh	868,429	453,560	414,869	487,796	302,371	185,425
Assam	23,248,994	11,983,157	11,265,837	14,327,540	8,324,077	6,003,463
Bihar	74,199,596	38,510,686	35,688,910	31,675,607	20,978,955	10,696,652
Chandigarh*	92,118	56,837	35,281	647,208	384,563	262,645
Chhatisgarh	16,620,627	8,290,983	8,329,644	11,283,183	6,770,898	4,512,285
Dadra and Nagar Haveli*	169,995	91,887	78,108	108,830	74,691	34,139
Daman and Diu*	100,740	63,576	37,164	111,939	72,559	39,380
Delhi*	963,215	533,219	429,996	9,703,049	5,713,157	3,989,892
Goa	675,129	339,626	335,503	989,362	544,006	445,356
Gujarat	31,697,615	16,289,423	15,408,192	29,050,019	17,349,179	11,700,840
Haryana	14,968,850	8,017,622	6,951,228	12,225,036	7,558,443	4,666,593
Himachal Pradesh	5,482,367	2,754,251	2,728,116	4,029,097	2,266,103	1,762,994
Jammu and Kashmir	7,564,608	3,925,846	3,638,762	4,704,252	2,999,353	1,704,899
Jharkhand	20,922,731	10,660,430	10,262,301	11,970,177	7,759,966	4,210,211
Karnataka	34,814,100	17,618,593	17,195,507	30,774,988	17,817,682	12,957,306
Kerala	23,571,484	11,450,785	12,120,699	25,625,698	12,817,963	12,807,735
Lakshadweep*	33,647	17,196	16,451	45,281	24,806	20,475
Madhya Pradesh	44,282,528	22,975,256	21,307,272	31,904,107	19,868,541	12,035,566
Maharashtra	55,732,513	28,443,238	27,289,275	64,566,781	37,487,129	27,079,652
Manipur	1,818,224	923,428	894,796	1,429,656	815,944	613,712
Meghalaya	1,853,457	939,803	913,654	1,170,443	619,274	551,169
Mizoram	450,018	233,718	216,300	663,262	351,851	311,411
Nagaland	1,635,815	846,651	789,164	1,146,523	645,807	500,716
Orissa	31,210,602	15,711,853	15,498,749	20,053,785	12,118,256	7,935,529
Pondicherry*	325,596	163,586	162,010	701,447	381,327	320,120
Punjab	16,043,730	8,500,647	7,543,083	14,853,810	8,515,310	6,338,500
Rajasthan	43,267,678	22,394,479	20,873,199	28,086,101	18,279,511	9,806,590
Sikkim	480,488	255,386	225,102	322,828	191,326	131,502
Tamil Nadu	34,869,286	17,508,985	17,360,301	40,624,398	22,847,735	17,776,663
Tripura	2,648,074	1,359,288	1,288,786	2,036,159	1,156,824	879,335
Uttar Pradesh	131,540,230	69,096,765	62,443,465	77,770,275	50,256,119	27,514,156
Uttaranchal	6,309,317	3,143,380	3,165,937	5,175,176	3,044,487	2,130,689
West Bengal	57,734,690	29,606,028	28,128,662	47,821,757	27,784,750	20,037,007

(continued)

TABLE 2.5. *(continued)*

India/State/ Union Territory*	% Change 1991–2001	Females per 1,000 Males	Total Population (%)			Urban (%)		
			Total	Males	Females	Total	Males	Females
INDIA	**21.34**	*933*	*100*	*51.73*	*48.27*	*27.78*	*52.61*	*47.39*
Andaman and Nicobar Islands*	**26.94**	846	100	54.17	45.83	32.67	55.11	44.89
Andhra Pradesh	**13.86**	978	100	50.56	49.44	27.08	50.89	49.11
Arunachal Pradesh	**26.21**	901	100	52.6	47.4	20.41	54.06	45.94
Assam	**18.85**	932	100	51.76	48.24	12.72	53.24	46.76
Bihar	**28.43**	921	100	52.07	47.93	10.47	53.5	46.5
Chandigarh*	**40.33**	773	100	56.41	43.59	89.78	55.81	44.19
Chhatisgarh	**18.06**	990	100	50.26	49.74	20.08	51.77	48.23
Dadra and Nagar Haveli*	**59.20**	811	100	55.22	44.78	22.89	59.15	40.85
Daman and Diu*	**55.59**	709	100	58.51	41.49	36.26	50.42	49.58
Delhi*	**46.31**	821	100	54.93	45.07	93.01	54.9	45.1
Goa	**14.89**	960	100	51.01	48.99	49.77	51.73	48.27
Gujarat	**22.48**	921	100	52.07	47.93	37.35	53.2	46.8
Haryana	**28.06**	861	100	53.73	46.27	29	54.14	45.86
Himachal Pradesh	**17.53**	970	100	50.77	49.23	9.79	55.64	44.36
Jammu and Kashmir	**29.04**	900	100	52.64	47.36	24.88	54.87	45.13
Jharkhand	**23.19**	941	100	51.51	48.49	22.25	53.47	46.53
Karnataka	**17.25**	964	100	50.93	49.07	33.98	51.55	48.45
Kerala	**9.42**	1,058	100	48.58	51.42	25.97	48.6	51.4
Lakshadweep*	**17.19**	947	100	51.35	48.65	44.47	51.66	48.34
Madhya Pradesh	**24.34**	920	100	52.09	47.91	26.67	52.67	47.33
Maharashtra	**22.57**	922	100	52.02	47.98	42.4	53.37	46.63
Manipur	**30.02**	978	100	50.55	49.45	23.88	49.77	50.23
Meghalaya	**29.94**	975	100	50.64	49.36	19.63	50.38	49.62
Mizoram	**29.18**	938	100	51.6	48.4	49.5	51.26	48.74
Nagaland	**64.41**	909	100	52.38	47.62	17.74	55.28	44.72
Orissa	**15.94**	972	100	50.71	49.29	14.97	52.77	47.23
Pondicherry*	**20.56**	1,001	100	49.98	50.02	66.57	49.85	50.15
Punjab	**19.76**	874	100	53.37	46.63	33.95	54.12	45.88
Rajasthan	**28.33**	922	100	52.03	47.97	23.38	52.91	47.09
Sikkim	**32.98**	875	100	53.32	46.68	11.1	54.71	45.29
Tamil Nadu	**11.19**	986	100	50.34	49.66	43.86	50.51	49.49
Tripura	**15.74**	950	100	51.27	48.73	17.02	50.98	49.02
Uttar Pradesh	**25.80**	898	100	52.67	47.33	20.78	53.23	46.77
Uttaranchal	**19.20**	964	100	50.9	49.1	25.59	54.05	45.95
West Bengal	**17.84**	934	100	51.72	48.28	28.03	52.84	47.16

(continued)

TABLE 2.5. *(continued)*

India/State/ Union Territory*	Rural (%)			Urban (%)		
	Total	Males	Females	Total	Males	Females
INDIA	*72.22*	*51.39*	*48.61*	*55.18*	*59.98*	*40.02*
Andaman and Nicobar Islands*	67.33	53.71	46.29	71	57.93	42.07
Andhra Pradesh	72.92	50.43	49.57	53.3	58.56	41.44
Arunachal Pradesh	79.59	52.23	47.77	44.71	61.99	38.01
Assam	87.28	51.54	48.46	53.79	58.1	41.9
Bihar	89.53	51.9	48.1	38.22	66.23	33.77
Chandigarh*	10.22	61.7	38.3	71.84	59.42	40.58
Chhatisgarh	79.92	49.88	50.12	54.26	60.01	39.99
Dadra and Nagar Haveli*	77.11	54.05	45.95	49.37	68.63	31.37
Daman and Diu*	63.74	63.11	36.89	70.82	64.82	35.18
Delhi*	6.99	55.36	44.64	70.4	58.88	41.12
Goa	50.23	50.31	49.69	73.61	54.99	45.01
Gujarat	62.65	51.39	48.61	57.41	59.72	40.28
Haryana	71	53.56	46.44	57.99	61.83	38.17
Himachal Pradesh	90.21	50.24	49.76	66.3	56.24	43.76
Jammu and Kashmir	75.12	51.9	48.1	46.72	63.76	36.24
Jharkhand	77.75	50.95	49.05	44.48	64.83	35.17
Karnataka	66.02	50.61	49.39	58.36	57.9	42.1
Kerala	74.03	48.58	51.42	80.49	50.02	49.98
Lakshadweep*	55.53	51.11	48.89	74.73	54.78	45.22
Madhya Pradesh	73.33	51.88	48.12	52.83	62.28	37.72
Maharashtra	57.6	51.04	48.96	66.73	58.06	41.94
Manipur	76.12	50.79	49.21	59.85	57.07	42.93
Meghalaya	80.37	50.71	49.29	50.75	52.91	47.09
Mizoram	50.5	51.94	48.06	74.44	53.05	46.95
Nagaland	82.26	51.76	48.24	57.65	56.33	43.67
Orissa	85.03	50.34	49.66	54.63	60.43	39.57
Pondicherry*	33.43	50.24	49.76	72.03	54.36	45.64
Punjab	66.05	52.98	47.02	61.15	57.33	42.67
Rajasthan	76.62	51.76	48.24	49.73	65.08	34.92
Sikkim	88.9	53.15	46.85	59.73	59.27	40.73
Tamil Nadu	56.14	50.21	49.79	65.41	56.24	43.76
Tripura	82.98	51.33	48.67	63.81	56.81	43.19
Uttar Pradesh	79.22	52.53	47.47	46.83	64.62	35.38
Uttaranchal	74.41	49.82	50.18	61.03	58.83	41.17
West Bengal	71.97	51.28	48.72	59.61	58.1	41.9

*Union territories, ruled directly by the national government.
Source: Registrar General and Census Commissioner, India (2001).

includes family planning, concern for the physical environment, and tolerance of other cultures. The Rishi Valley experiment exemplifies the attempt to reverse the course of destruction and offer alternatives to emerging consumption patterns, and it represents an important step for Indians and members of other cultural groups who seek to alter the course of environmental destruction through embracing the concept of sustainability. Rishi Valley educators are not attempting to set aside spaces entirely free of development. On the contrary, they attempt to introduce the type of corrective measures, land-use practices, and environmental programs that improve the quality of life for the people of a region, without denying future generations their chance to have access to environmental resources and a reasonable quality of life.

We explore issues of sustainability in greater detail later in this volume. Our point here is that population growth, consumption, and the environment are intricately connected, which should give us hope that sustainability will help maintain the delicate balance. Sustainability does not promote a blind return to some normative nature, but rather embraces the notion of the human need for progress and an improved quality of life. Its appeal is its balancing of the needs of both future and present. It also extends to cultural, social, political, and economic structures to ensure that inequities are understood and addressed within its design.[8] Globally, this means that sustainable development needs to address the imbalances between countries in the North and South. For example, why should countries in the South reduce their population and be blamed for impending demographic apocalypse, when countries in the North continually expand their energy needs? This is not a theoretical question but one that has been expressed forcefully in various global gatherings. Policy initiatives on population control have squared off against the political economy of development in the last few meetings of nations, during which the West, especially the United States, has faced vociferous groups from developing countries denouncing a one-sided perspective on the environment.

SOMETHING NOT SO FUNNY HAPPENED ON THE WAY TO CAIRO

> Carbon monoxide and sulphuric acid emissions are another major technologically induced problem affecting air quality, especially in major cities . . . worldwide atmospheric concentration of carbon dioxide (CO_2), the principal greenhouse gas, has gone up by 9%.

[8]For a recent edited volume that addresses these issues, see Agyeman et al. (2003).

The developed economies, with slightly over 15% of the global population are and have been the major contributors to carbon emissions, with over 46% of total output. The major sources are the burning of fossil fuels and cement manufacturing, both strongly correlated to urbanization and industrialization. It was estimated that in 1989, for every million dollars of GDP, an additional 327 tons of carbon was released into the atmosphere. On a per capita basis, this meant an average of 0.5 tons for the low and middle-income regions of the planet and 3.36 tons in the advanced economies. The world average is 1.12 tons per capita. When population is factored in, industrial economies produce over six times the per capita carbon dioxide pollution of the rest of the world. However, as industrialization and urbanization expand in poorer regions, pollution there has tended to increase at a faster rate than in more developed areas. While Europe, North America, Japan and other industrialized economies, although generating a much larger amount of CO_2, are increasing air contamination by roughly 0.5% per year, the less-developed economies are doing so at an annual rate of 3.8%: 7.6 times faster. (Nef 1995)

This excerpt from Nef's book *Human Security and Mutual Vulnerability* is one account of global imbalances in the production of pollution. While Nef points to urbanization and industrialization as the processes most responsible for the rapid production of pollutants, he highlights two themes that remain contentious in population debates. First, developed countries produce a disproportionate level of the air pollutants. Second, developing countries are increasing their pollution at a much higher rate than developed countries. This means that as less developed countries become urbanized and industrialized, they will arrive at a per capita pollution production rate equal to or exceeding that of the West. What's missing from this debate, however, is the third factor. To what extent is production in the developing countries aimed at consumption in developed countries?

Regardless of political posturing, we reach an impasse when issues of population control and sustainability are given global urgency. While countries in the South are asked to reduce their population, countries in the North are asked to reduce their pollution. But both need also to control their consumption of resources.

This policy dilemma has manifested itself in a number of ways at every global summit since Bucharest in 1974. To illustrate some of the actual debates on this topic, let us consider the United Nations report on the decennial meetings.

Following two preliminary and largely technical meetings on population issues in Rome (1954) and Belgrade (1965), the World Population

Conference in Bucharest (1974) represented the first intergovernmental conference on population, and it adopted the World Population Plan of Action (United Nations Economic and Social Council 1995). This meeting was followed by one in Mexico City (1984) and another in Cairo (1994). At all of these meetings, participants discussed population within environmental, economic, and social contexts. Whereas in Bucharest, however, development was seen as a cure for many social ills, including uncontrolled population growth, delegates in Cairo had become more sophisticated. Based on twenty years of experience since their first Plan of Action, the Programme of Action formulated in Cairo addressed a number of interesting issues. These included human rights, respect for indigenous cultures and their perspectives on population, women's rights, urbanization forces, production and consumption patterns, and, most important, sustainable development. By 1994 most countries had established some form of family planning policies. Many had also become aware that development by itself was not a magical solution to rapid population growth. Connection between social, political, economic, environmental, and cultural factors had become better understood. Participants in the Cairo conference thus reaffirmed the agreements in Rio and in the provisions of Agenda 21. In other words, the more developed countries of the West, especially the United States, were put squarely in the middle of population discussions. Whereas Bucharest's demand for development matched the emerging belief among the intellectual elites of the West that development was a precursor to democracy (the seminal work of Lipset [1959] had a profound influence here), and hence of socioeconomic improvements under a free market economy, by 1994 discrete discussions of population, environment, and economy were tied together in a discourse on sustainability. In the 1960s and the 1970s Western nations were happy to aid development programs that brought funding along with population control requirements to the developing world. By the 1990s, however, developing countries wanted more sustainable development at home as well as in the developed countries of the West. Furthermore, women's rights became a central part of the population debate.

At the June 1992 Rio Summit, three groups with three distinct—albeit interconnected—interests came together to create a volatile condition for policy debates.[9] Even though the Rio gathering was to focus on the

---

[9]See Cohen (1993) for a brief analysis of what happened at Rio. This paragraph relies on some highlights from this paper.

environment, the three issues of environment, population, and women became the main topics of debate. The meeting began with an appeal by a number of delegates from countries in the global South to deemphasize population growth as a primary cause of environmental problems. To them, equating the two things would put the burden of environmental correction on the countries in the South and would ignore issues of overconsumption in the North. At a meeting in New York, where the so-called Agenda 21 was to be drafted prior to the gathering in Rio, things got quite contentious. While the U.S. delegation objected to any reference to quantification of consumption in developed nations, the nonaligned countries (Group 77), led by Pakistan, retaliated by attempting to weaken the chapter on population. Ironically, the stance on population issues brought the Vatican and some women's rights groups to the side of the nonaligned countries. In the end, the document was weakened on all fronts, including population issues, which ensured that almost everyone would be unhappy with it.[10] All was not lost, however. Sustainability and population were clearly connected, as were issues surrounding women's rights, especially women's control of their own reproductive systems.

Although the 1994 Cairo International Conference on Population proved contentious, delegates arrived at a surprisingly positive consensus and moved further than they had in Rio. Instead of focusing on population growth alone, they brought issues of individual rights in the area of sexuality and reproduction to the center of their debates (DeJong 2000). As a result, by the mid-1990s cultural, social, and economic factors had become part of discussions of population and the environment. It was no longer enough to discuss lowering fertility rates in nations of the South as an end-all solution to poverty, food security, environmental degradation, and other problems.

In addition, during the 1990s a number of publications on the implementation of Agenda 21 in various countries began to highlight the importance of sustainability, which complicated the issue of population in a positive way. Some researchers, such as Briffett et al. (2003), began to suggest that Agenda 21 was creating positive momentum toward strategic environmental assessment in the developing nations of Asia and increasing environmental awareness among top-level personnel of various governments. Other researchers (Spangenberg and Lorek 2002; Spangenberg et al. 2002; Valentin and Spangenberg, 2000) criticize the

---

[10]Selected sections of Agenda 21 and the Rio Declaration can be found in Delegates of the United Nations Conference on Environment and Development (1992).

muddled way in which Agenda 21 articulates institutional roles and its absence of a structured and measurable indicator of the influence of various agents (e.g., households) in creating environmental pressures. These critics presented the beginning of a model for understanding environmental pressures of households through housing, eating, and mobility patterns.

Rotheroe et al. (2003) extend the principles of sustainability in the Local Agenda 21 to the business community. They suggest that neglecting the role of individual enterprises will create unsatisfactory results and they have developed a model that would incorporate such enterprises into Agenda 21. Interestingly, despite the heavy political posturing before, during, and after the Rio Summit, it appears that discussions of sustainability, albeit interpreted and implemented unevenly, have informed the planning processes of a number of countries. As these discussions continue, the role of various stakeholders is identified. For example, in many cases we are beginning to see a stronger emphasis on collaboration between NGOs and governments in establishing sustainable agendas (see, for example, Steinberg and Sara 2000 and Briffett et al. 2003; Roddick 1997 suggests that the trend toward including NGOs in discussions of sustainability began in Rio and is embedded in Agenda 21).

As sites of overconsumption, cities must be seen as places where both population growth and energy/resource consumption must also be controlled or curtailed. The Local Agenda 21 has created a more rigorous discourse on sustainable development in the cities of both North and South.[11] As cities begin to assess their energy consumption, land-use patterns, hazardous sites, air pollution, traffic, food shortages, poverty, and homelessness, they appear to be arriving at similar conclusions regarding their future.

Unfortunately, these positive trends are overshadowed by a set of new concerns. For example, as bureaucrats educate themselves and their constituencies about what a sustainable city may look like—as the concept of sustainability makes its way into vernacular discourse— the post-Rio era presents us with some fundamental questions about the assumptions of that summit. Are economic growth and environmental protection compatible, or is sustainability an impossible

---

[11]Since 1992 hundreds of scholarly articles and public documents have been written on Agenda 21 and sustainable development. McMahon (2002), Bennett and Newborough (2001), Tominaga (2001), Rahardjo (2000), Otterpohl et al. (1999), and Laituri (1996) offer examples from various regions of the world.

goal?[12] Certainly, two-dimensional discussions of population have been displaced by a more systematic concern for the global community. Eco-cities have become the latest concept in the emerging discourse[13] and international "soft-law" treaties have also come to occupy legal international space vis-à-vis environment, population, social equity, political stability, economy, development, and poverty. Are we seeing here a new challenge to established notions of sovereignty and how it may play out in the emerging eco-politics?[14] As countries in the North see the discussion on population shift to sustainable development, the requirements of environmental stewardship demand contributions from every country, whether in the form of old-fashioned fertility reduction, lowering air pollution, or limiting consumption. Just as the political and economic restructuring of the world in the nineteenth and twentieth centuries resulted in the creation of nation-states with rigid boundaries, it is possible that environmental concerns are becoming the collective means of undoing these artificial borders and calling into question the very meaning of nation-states. Will the legacy of the twenty-first century be a new way of looking at national boundaries and the governance of nations?

Two centuries after Malthus, while we remain aware of the looming disaster of unchecked population growth, we face a radically transformed world. As Engels, Marx, and Ricardo argued, the population problem is embedded in a class-based society with uneven patterns of development, property ownership, and resource distribution (Foster 2000). As capitalism has moved into its advanced phase, corrections to population control will necessarily be economic and political in context. Given that market forces are unlikely to provide a global pattern of social and economic equality anytime soon, the issue of population control remains a contentious one. When Malthus was writing, areas of the world were suffering from colonialism, and many of today's nation-states had not yet been born. The birth of modern states further complicated the issues of interregional connectivity and contributed to uneven development. States at different stages of development have differing concerns with population, environment, and sustainability. Finally, given that the mobility of capital in this era of globalization has created a planetary division of labor and increased the gap between

[12]An example of such a discussion is provided in Clark (1995).
[13]For a discussion of the concept of the eco-city and its relationship with sustainability, see Roseland (1997).
[14]For discussion, see Litfin (1997).

rich and poor, international treaties, vis-à-vis eco-politics, may be a force for diminishing sovereignty and creating a world of multinational corporations. Ironically, as the hegemonic forces of capital diminish the importance of national borders, global awareness of their costs has increased. A new discourse on individuality, human rights, and the ability to determine one's future has arisen to counter the hegemony of capital. This view has turned the population discourse and its Malthusian parentage on its head. Does the policy debate on population continue to be normative (see MacKellar 1997)? Shouldn't respect for individualism and women's rights in controlling their own reproductive systems encourage us to shift the debate to other topics such as equity and fairness? Shouldn't we bring environmental justice to the center of our discourse on population, environment, and sustainable development? The trajectory of an urban future suggests that population and environment are a part of the sociopolitical, economic, and cultural domain. We can surely accept that unabated population growth will spell disaster for all of us, but who produces, who consumes, and who inherits the burden of development are complicated questions that can destabilize our societies before the earth can reach its imaginary carrying capacity. The demons of nationalism and capitalism have created many of the problems we face and surely cannot be the solutions to them. Our imagined sustainable future may depend on how successfully we can exorcise them from our lives.

3 Feeding Cities That Consume Farmland

URBAN SPRAWL is derided by critics as wasteful. Low-density suburban development generally leads to an inefficient use of resources. Services such as water and sewers are more expensive because longer lines serve fewer people than in densely settled cities. Public transit is inefficient for the same reasons—traveling longer distances through lower-density markets means higher costs and fewer passengers, all of which requires a high subsidy from taxpayers or from transit fare payers in the city. Curvilinear street patterns with few intersections, typical of suburban developments, are also not amenable to public transit. For rail transit, building straight lines is easier and less expensive than building curved ones. Long distances between intersections and natural pick-up and drop-off points make all forms of transit less efficient in suburban areas. And because suburban developments often create large residential developments, distances to destinations—such as shopping or work—tend to be longer than in downtown cores. All of these factors encourage the use of private vehicles, which consume vast amounts of energy and move people very inefficiently. In short, suburbs consume high amounts of energy (Khisty and Ayvalik 2003).

Increased reliance on the automobile and inefficient use of energy and municipal services are not the main factors that capture the public's attention when it comes to urban sprawl. Encroachment onto farmland, however, tends to raise concern. The conversion of pastures into parking lots or forests into subdivisions while central city properties lie vacant or underused is an all-too-familiar situation for many municipalities. The American Farmland Trust estimates that every year in the United States, 1.2 million acres of productive farmland are converted to developed land uses (American Farmland Trust n.d.).

While the number appears large, cities occupy surprisingly little territory. In the United States, for example, urban areas occupy only about 3 percent of the nation's land. The pressing issue is not the quantity of land being converted to urban use but the amount of good-*quality* farmland. The U.S. Department of Agriculture estimates that in the United States, prime farmland is being converted to development at two to four times the rate of less productive farmland (USDA 2002a). Another study of farmland loss shows that the overall quantity of cropland in the United States has remained fairly constant over the past two decades, but this is largely the result of converting rangeland in the semiarid west into cropland, which is often irrigated. Such a shift can enhance plant productivity and profit, but many question the sustainability of irrigated agriculture compared to the rain-fed agriculture in the humid East that it is replacing (Greene and Stager 2001).

It is no accident that urban areas sit on top of some of the best farmland in the world. The richness of the soil is what brought people to these areas in the first place. Even in wealthy countries where relatively few people are engaged directly in farming, major cities are often found, and growing, in areas with good soil. Although the wealth of modern cities is no longer dependent solely on the bounty of agricultural hinterlands, the early advantages to cities of rich soil for farmers' markets has kept many of the cities in the forefront of their urban hierarchies (Pred 1980). When we understand that most cities began as farmers' markets, we can begin to comprehend the paradox that cities consume the farms that feed them.

AGRICULTURAL ORIGINS: CITIES AS FARMER'S MARKETS

The first cities, like those today, were gathering places for specific human purposes. Some ancient cities were planned and constructed for religious reasons, or as centers of kingdoms for control and regulation of their domains. Other cities emerged organically as unplanned centers of trade. Whether planned or unplanned, cities depended on the surplus production of food. People could live in cities only if others grew food for them, a fact that is as true today as it was ten thousand years ago. Surplus food was produced in regions of agricultural bounty, areas with abundant sunshine, soil nutrients, and water. It was in these fertile areas that farmers were able to trade their surpluses at temporary, and later permanent, markets for other food and eventually other goods and services. The first permanent urban dwellers facilitated the trade of food or offered other products, such as clothing, in exchange for food. Again, little has changed for urban dwellers today, who must still offer some good or service in exchange for the food that sustains them.

In the United States, farmers make up only 2 percent of the labor force, down from 38 percent a century ago. As recently as 1950, one in nine persons in the labor force was employed on a farm (U.S. Census Bureau 2004). A similar decline in the number of farmers is evident in other wealthy countries, where increasing numbers of people are dependent on fewer people to grow food for them. In countries with advanced economies, about three-quarters of their populations live in cities. Urban residents depend on the ability of relatively few people to grow food in huge quantities. Many also depend on agriculture for employment, even if they are not farmers. In the United States, food-related industries and services account for 25 percent of all jobs. Annually, agricultural products contribute more than $150 billion to the GDP. Agricultural value added in the European Union amounted to $155 billion in 2001 (World Bank 2004). Given the importance of agriculture to national economies, as well as concerns about food security and the strength of powerful agricultural lobbies, most countries heavily subsidize agricultural production. In other words, even though wealthy countries employ few farmers, many jobs and the economy in general, including in urban areas, are based on agricultural productivity.

FROM HINTERLAND TO ECOLOGICAL FOOTPRINT

Historically, cities drew principally upon immediate territory for resources, including food. Cities tended to grow on rich agricultural soil because the productivity of the soil created more opportunities for trade, the economic basis of the markets that defined urban areas. While city dwellers may have traded for goods from distant places, most bulky items, including foodstuffs, were far too expensive to ship great distances.

One of the remarkable changes in the twentieth century was the rapid reduction in transportation costs. Prior to the development of the railroad, transporting goods over land was difficult and costly. As a case in point, in 1800 it was cheaper for merchants in Pittsburgh to ship goods down the Ohio and Mississippi rivers through the Gulf of Mexico and along the eastern seaboard than to transport it three hundred miles overland to Philadelphia. Given that riverboats charged two cents per ton-mile, while wagon transport cost fifteen cents per ton-mile, farmers could not make a profit if they had to ship their goods more than a hundred miles overland (Earle 1990). Because of the high costs of transportation, markets had to rely on the produce of their hinterlands, or surrounding area. Canal building in the 1820s and 1830s, followed by the railway construction boom of the subsequent decades, drastically reduced transportation costs, allowing goods, including bulky food items like wheat, to travel farther, thus extending the hinterlands of markets. The staggering success of Chicago depended on the cheap transportation provided by the railway and through the Great Lakes that permitted the city to draw on the enormous and rich agricultural hinterland of the U.S. Midwest (Cronon 1991). By the 1880s, transportation rates had dropped to less than a cent per ton-mile (Earle 1990).

Transportation technologies, along with large sums of capital, transformed the American landscape by "commodifying nature." Manipulating the natural processes of plant reproduction for human gain increased the reach of human domination of the landscape. Sophisticated trading mechanisms like the Chicago Board of Trade amplified the ability of people to turn forests and prairies into farmed fields, opening up rich agricultural lands that could feed growing cities in eastern markets (Cronon 1991). The industrialization of the economy, in other words, resulted in a vast expansion of agricultural hinterlands, with very real environmental consequences. Besides the deforestation of the American landscape, heavier reliance on mechanized forms of transportation marked the beginning of the age of fossil fuel. Although wood, given its natural abundance in North America, continued to be an important source of energy in the nineteenth century, by the 1880s coal was the dominant energy source in nearly all U.S. cities (Williams 1990). The transition from horse and cart to railroad and coal meant improved efficiencies in economic terms but began a chapter in the overexploitation of fossil fuels that the world continues to deal with today (Buckley 2004).

Any casual observer at a supermarket will notice that food comes from around the world. Previously unimaginable efficiencies in transportation, coupled with low farming costs (both labor and machinery), mean that products can be shipped thousands of kilometers to market and still make a profit, unevenly distributed though it may be. Under these conditions, the idea of a hinterland becomes an abstraction. It is difficult for consumers today to see the direct connections between their consumption and its impact on the land, for the hinterland is no longer the territory beyond the city limits but encompasses the entire globe.

The concept of the ecological footprint attempts to overcome the limitations of thinking in terms of market and hinterland. By measuring the amount of biologically productive land and sea required to support human actions, this ecological accounting system draws a direct link between consumption and waste and the resources necessary to support them. It is possible to calculate an ecological footprint for an individual or group, depending on the use of resources and production of waste. The United States, given its heavy use of energy and high rates of consumption, has one of the largest ecological footprints in the world, requiring 10.3 hectares, or about twenty-five acres, of biologically productive land or sea per capita to support its way of life. At current population levels, the world has approximately 2.3 hectares of biologically productive land and sea per person. By this calculation, many countries have ecological footprints that far exceed the carrying capacity of their territories. Italy, for example, has an ecological footprint that is more than three times the total area of the country (Wackernagel and Rees 1996).

The contrast between ecological footprint and area occupied is most apparent in cities. Although urban areas occupy relatively little space on the earth's surface, their ecological footprints are very large, as vast amounts of energy are consumed and huge amounts of waste are produced in these agglomerations. Columbus, Ohio, for example, is a medium-size city of 1 million people that occupies most of Franklin County, an area of 140,000 hectares. Assuming Columbus residents match the national average of 10.3 hectares per person, the amount of territory necessary to provide Columbus with the resources it needs and to absorb its wastes is approximately 10 million hectares, about seventy times the area the city occupies. Using this formula, Columbus has an ecological footprint almost as large as the entire state of Ohio. Given that other people reside in the state, including the larger metropolitan areas of Cleveland and Cincinnati, Ohioans must rely on territory beyond their borders, including territory overseas, for resources and waste absorption. The rainforests of the Amazon River basin, for example, help to absorb some of the carbon dioxide released from coal-generated electricity plants in Ohio. The same area provides tropical fruits, wood products, beef, and other agricultural products to Columbus. In other words, wealthy countries are reaching beyond their boundaries to sustain their current ways of life, resulting in an uneven global distribution of resources.

The ecological footprint concept has broad currency because it gets around the problem of hinterland and measures the global impact of human activities. It is a system that can work at a variety of scales, from individual to global. It does, nevertheless, have limitations. Some have criticized its utility as a measurement tool, arguing that the system is not an accurate metric of human impacts but only an approximation, and a poor one at that (Herendeen 2000; van Kooten and Bulte 2000). Yields, for example, are used as measurements of productivity for arable land but do not take into account the economic factors that explain differences in yields. Given similar economic and market structures, yields in poor countries may be considerably higher. If yields could be raised, this could potentially reduce pressure on unfarmed land while at the same time increasing the carrying capacity of the planet (van Kooten and Bulte 2000).

A more general critique of the ecological footprint is that it does not address, from the point of view of sustainability, indices of human welfare. Human activities that may be improving the lot of people through poverty reduction might also be creating larger ecological footprints. Yet, by reducing poverty and the concomitant "pollution of poverty" dilemma, a temporarily expanded ecological footprint might create the wealth and knowledge to reduce the future ecological footprint of the planet (Opschoor 2000; World Commission on Environment and Development 1987). As well, some parts of the planet may be more adept than others at providing ecological services. The boreal forests of Canada may be better suited to absorbing carbon, while the midwestern United States may be better at growing food. By expressing ecological footprints in terms of deficits, the concept implies that countries or regions should live within their ecological means rather than encourage the trading of ecological services (Ayres 2000).

Finally, the ecological footprint is a static tool, not designed to incorporate human and ecological dynamics, which limits its application for planning and policy. For these and other reasons, critics view the ecological footprint concept as a neat conceptual tool for linking human behaviors with global resources but consider it unworkable as a policy instrument.

At the same time, even critics admit that the ecological footprint has been very successful in capturing the public's attention. While it may not be a perfect accounting system, it draws a direct link between consumption and land, a tangible and finite product (Rees 2000; van Kooten and Bulte 2000). The footprint is a particularly potent tool for urban residents because it illustrates the extent of a city's impact on the environment as well as its dependence on the environment for survival. While cities occupy little territory, one of their general characteristics, particularly in wealthy countries, is rapid growth on the periphery. Both the physical and ecological footprints tend to expand with urban sprawl, as energy demands tend to increase with the typical automobile-dependent lifestyles that characterize low-density suburban developments.

URBAN SPRAWL AND FARMLAND LOSS

The urban-rural fringe area is where the infamous sprawl battles take place. What makes them battles is that most people concerned with the negative consequences of sprawl point to farmland loss as the major problem, while for developers it is the "greenfield" sites, already cleared, that make them prime locations for construction (Feather and Barnard 2003). For the most part, the developers are winning. Despite the efforts of governments and nonprofit organizations, suburban sprawl continues to accelerate in most wealthy countries, particularly in North America. Nevertheless, many groups have made small strides in protecting farmland from the bulldozers.

The preservation of farmland has a relatively long history in western Europe, where population densities and food security issues have created perceived threats to agricultural resources (see the section "Urban Growth Boundaries and Greenbelts" in Chapter 6). In North America, farmland conservation has a shorter history. In part this is related to geography. Because it is a vast continent with rich agricultural soil able to produce large surpluses of food, placing limits on the conversion of farmland would have struck most people as absurd until relatively recently. Indeed, in many parts of the continent today, such an idea still brings a smile of bewilderment. But in the large urban conglomerations, such as the corridors between Boston and Washington and San Diego and San Francisco, and the Toronto to Buffalo conurbation, farmland preservation is more likely to find a receptive audience.

Why do people in North America want to preserve farmland? One notable scholar (Bunce 1998) argues that there are two ideological movements—environmentalism and agrarianism—that account for the recent interest in farmland preservation. The rise of environmentalism in the 1960s reflected increased awareness of the relation between people and the land that sustains them, and coincided with concerns for farmland preservation. Malthusian treatises on population, such as Ehrlich's *Population Bomb* (1968), created a sense of anxiety about the ability of the world's arable land to feed rapidly growing populations. Joni Mitchell's famous lyric—"they paved paradise, and they put up a parking lot"—captured the feelings of many North Americans, who began to question the wisdom of unrestrained growth and its effects on a critical and finite resource. When dire predictions of the massive loss of farmland did not come to pass, farm preservationists turned to environmental ethics in the early 1980s, arguing that conserving farmland was one way to achieve ecological stewardship of the land. Tied into the conservation ethic is a strong undercurrent of agrarianism (and antiurbanism) that permeate North American culture. Beyond the Jeffersonian ideal of the yeoman farmer and civil society, farms, especially family farms, embody a romantic nostalgia for the rural origins of North America (Bunce 1994; Lehman 1995). In many suburban and exurban developments, that "ruralness" is advertised as an amenity. Old farmhouses are often retained on new developments as a way of commodifying an idealized rural heritage. Working farms are encouraged through farmland preservation because they bring authenticity to the rural experience (Bunce 1998).

In California farmland preservation programs have been relatively popular, especially in the San Francisco Bay area and along the coast. Yet only fifty thousand hectares, or less than half of a percent of the 11 million total agricultural hectares in the state, have been protected through easements (Sokolow 2002). Although the numbers are small, most easements have been purchased on the urban-rural fringe, where farmland conversion is most likely to take place. If planned with care, the easements can discourage "leapfrogging," or the development of residential areas beyond the built-up areas of cities (Meadows 2002). On the other hand, if easements are purchased in small pieces and in an uncoordinated fashion, the process can encourage leapfrogging, arguably making urban sprawl worse than growth on the periphery of existing cities.

The California State Conservation Program, established in 1996, provides funds to conserve farmland by purchasing easements. Local governments may also restrict farmland conversion through zoning, while nonprofits and other groups can purchase easements as well. A recent study of land trusts in California showed that environmental concerns and protecting agricultural resources are top priorities (Sokolow and Lemp 2002). Some land trusts see their role as simply preserving open space, promoting "a landscape free from human congestion and an antidote to urbanization" (Sokolow and Lemp 2002, 12). These concerns bear out Bunce's (1998) review of the North American farm preservation movement as a combination of environmental and agrarian ideologies.

On the other side of the equation are the farmers. Buyers require sellers, and the aspirations of both parties may differ, even if the end result is an agreement to give up development rights. Farmers sell develop-ment rights for a number of reasons, and income is a major induce-ment. In three counties in the Greater San Francisco area, farmers are paid hundreds of thousands of dollars, and in some cases millions, if they agree to keep their land as farms in perpetuity. Another reason for selling development rights is that it provides income that allows farm-ers to continue farming on land to which they have a sentimental attach-ment, land that has been in their families for generations and that they may want to pass on to the next generation. In other words, easements provide a form of security in terms of both income and retaining the land as a farm (Rilla 2002). Economic rationale alone does not explain the behavior of farmers who participate in easement programs.

The growth in farmers' markets in the United States, from 340 in 1970 to more than three thousand in 2001, offers another rationale for preserving farmland on the periphery of cities (Brown 2002). Since most farmers' markets sell only goods that farmers produce themselves, there is a market for locally produced food, which helps to sustain farming on the urban fringe. Many farmers survive by tapping into these niche markets, although many supplement their farming income with other forms of work. Organic and local produce can often fetch high prices, which some consumers are willing to pay because of real or perceived higher quality and because they want to support local farmers. Con-sumers also value farmers' markets as lively, fun places to be and are supported for the pleasure they bring in this regard. For environmen-tally minded consumers, the fact that the food comes from nearby is important because it means that less fossil fuel energy was consumed in shipping the goods to market and because it encourages agricultural preservation (Brown 2003).

Farmers' markets provide other community and personal benefits. Face-to-face contact between food producers and consumers lessens the abstraction of food consumption by connecting consumers with the people who grow food. Farmers' markets have also been encour-aged as a means of assuring food security, especially in lower-income neighborhoods, where inadequate nutrition remains a problem. Recog-nizing the nutritional value of farmers' markets in poor neighborhoods, the USDA Farmers Market Nutrition Program provides WIC coupons to pregnant and nursing mothers to use at farmers markets (Gottlieb and Fisher 1995). In 2002, more than 2 million people received coupons for produce that could be purchased at farmers' markets. This pro-gram's success resulted in a new program in 2003 to provide seniors with similar benefits (USDA 2003). While farmers' markets in most North American cities are unable to meet all the nutritional needs of all city dwellers, they are one way to encourage farmland preservation.

A threat to farmland preservation efforts lies in the difficulty of farming on the urban-rural fringe. Given the high cost of land, farmers on the periphery of cities grow intensively, trying to pull from the land as much as they can to offset the high cost of property. Truck farming or market gardening is a typical form of agriculture on the urban fringe, as is dairying. High-value items are often supported through the heavy application of fertilizers and pesticides. Dairy farms can be very smelly. Heavy machinery is noisy and pollutes the air. While the farms may be pretty, many new homeowners are chagrined by the reality of a working farm (Bunce 1994). Such conflicts of interest can pressure farmers to sell.

URBANIZATION AND AGRICULTURAL INTENSIFICATION

One controversial approach to curbing farmland loss is bioengineering. Through the genetic manipulation of crops, scientists can increase yields while providing plants with the ability to fend off pests and disease without the use, or with reduced use, of pesticides and herbicides. If yields can be increased through genetic engineering, it is possible that less of the world's arable land will need to be used for agriculture and more can be devoted to other uses, including bioreserves. Some advocates have even suggested, contrary to expectations, that genetic engineering, which promotes monocropping, can actually increase biodiversity, assuming that the higher yields will reduce pressure on farmlands and allow more land to revert to natural cover (McClintock 1999).

The promises of bioengineering will probably have the same impact as the green revolution of the 1950s, and arguably even of agriculture, when it was first invented ten thousand years ago. That is, it will probably continue to push people out of rural areas and draw them to cities. Human beings, by controlling the growth of plants for fiber and food, have been able to assure their survival and rapid population growth. Despite the dire warnings of Malthus, we have been able to grow enough food to feed our rapidly expanding population over the past century (although there have been times of regional deficits). This enormous productivity has been made possible primarily through the intensification of agriculture, or getting more food from existing arable land. Malthus did not foresee three things—that people could be willing to control family size, that they would have an economic incentive to do so as they moved to cities and participated in an industrial economy (where child labor was valued less than on the farm), and that technologies could increase food production faster than the exponential population growth he predicted.

Development of inexpensive inorganic chemicals, especially nitrogen, in the nineteenth century boosted farmers' yields considerably. Increases in productivity were initially limited to parts of the world that could afford to buy the synthesized fertilizers. Rich countries with cash economies first took advantage. Improved breeds of plants also allowed farmers to get more from the same amount of land. For millennia, humans have selectively improved food plants. Maize, commonly known as corn in North America, originally was the size of a human finger but through selective breeding reached its current size, long before the arrival of Europeans in the New World. Over the course of thousands of years, individuals have modified plants from their natural state in order to improve yields, taste, storability, and other characteristics. Adding cheap inorganic fertilizers helped to boost yields of those plants already modified by humans over hundreds of generations. Surplus production of food allowed farmers to purchase machinery and more fertilizer, both resulting in a reduced need for human labor. Sons, daughters, and itinerant laborers with scythes were replaced by mechanical reapers, reducing the demand for labor on the farms. Without jobs on the farms, cities became the destination for such people. At the same time, urban dwellers required others to grow food for them, which is what the machinery and fertilizers allowed. In exchange, city dwellers produced other goods, including farm machinery, in the factories that popped up in industrial cities in the late nineteenth century. Urbanization and the "industrialization" of agriculture in this way went hand in hand.

The green revolution of the 1950s answered a need to ramp up food production, especially in poor countries where population growth was occurring fastest. Modification of staple crops, particularly rice, wheat, and maize, vastly increased yields and also reduced growing times. Traditional rice varieties, for example, take 150 to 180 days to reach maturity, while newer varieties mature in as few as 100 days, permitting double- and sometimes triple-cropping (Davies 2003). Between 1961 and 1980 food production increased by 3.2 percent per year in all developing countries; from 1981 to 2000 it increased 2.2 percent. Part of the increase was the result of increased area (extensification), but the majority (2.5 percent and 1.8 percent, respectively, for these periods) was the result of increased yields (intensification) (Evenson and Gollin 2003).

Critics of the green revolution are quick to point out that its benefits have been uneven. Despite rapid growth in food production, nearly 800 million people remain undernourished (FAO 2004). To participate, farmers require sufficient funds to purchase hybridized seeds and chemical fertilizers. Large landholders benefited most from the high-yield varieties, especially in the early years. But diffusion of high-yield varieties into the hands of smallholders has spread the benefits more widely in recent decades (Davies 2003). A second charge is that the green revolution has introduced chemical farming into poor countries, leading some supporters of biotechnology to call for a "greener" green revolution with less reliance on chemical inputs (Conway 1998; Conway 2000). One of the promises of genetically modified food crops is that disease and pest resistance can be engineered into the plants, thereby reducing the need for pesticides and other forms of pest management. Genetically modified food crops, however, have met resistance, especially in Europe, as consumers are uncertain about the impacts of such crops on their health as well as on ecosystems.

There is little question that the green revolution set the world on a trajectory of increased food production. In wealthy countries, yields continue to rise. The USDA, for example, projects that yields of major crops in the United States will rise significantly over the next decade. Projections for corn are from 137 bushels per acre in 2003 to 153 in 2011, a 12 percent increase. Total production of corn is expected to jump from 9.9 to 11.2 billion tons over the same time period (USDA 2002b).

Along with increased production have come reductions in the cost of food, although the benefits have been uneven. Relative to income, food costs have been declining in most wealthy countries. Individuals in the United States, for example, spend a small fraction of their income (about 10 percent) on food. In poor countries, that figure can be as high as 70 percent (World Commission on Environment and Development 1987). While incomes tend to be higher in urban areas than in rural ones, the urban poor must spend a large amount of their income on food. This has led many city dwellers to search for other ways to obtain food, including farming in cities.

URBAN AGRICULTURE

Cities have a long history of being defined as nonagricultural. This urban-rural dichotomy is a convenient yet oversimplified division of human activities. For millennia people have raised livestock or planted vegetable gardens in cities. Sometimes livestock roamed freely in city streets or grazed in common areas (the Boston Common was originally designed for that purpose). Produce could be raised in backyards or community gardens, found in many cities to this day. While raising livestock has been made illegal in the cities of most rich countries, the clucking of chickens can still be heard there, especially in poor immigrant communities. It is estimated that 800 million people worldwide practice urban agriculture (United Nations Development Programme 1996).

Defining urban agriculture is complicated by the fuzzy and imprecise definitions of "urban" (see Chapter 1). More agricultural activity certainly occurs on the boundaries (fuzzy themselves) of urban conglomerations than in densely built-up cores. The International Development Research Centre (IDRC) defines urban agriculture as "a practice located within or on the fringe of a town, a city or a metropolis, which uses human and material resources to grow, process, and distribute a diversity of food and non-food products to those intra and periurban areas on a daily basis" (IDRC 2003). Many development organizations include agriculture on the urban fringe, or "periurban" zones, as part of the definition. The difficulty with this definition is that it does not define a town, city, or metropolis, or the extent of periurban areas. If urban agriculture is to be treated as something extraordinary, or as different from typical farming, it should refer to farming activity within the *built-up* areas of cities, in places where one would not expect farming to occur. In this sense, we can define urban agriculture as farming in densely built-up cities where farming is not typically practiced.

Normally farming does not occur in high-density urban areas because the land is very valuable and is best suited for other purposes. Von Thunen's explorations into agricultural land use made the arguments very clear. The bid-rent curves from the urban core help to define and delineate land use (Figure 3.1). While an elegant theory, the reality is different, as the bid-rent curve is interrupted by natural features, such as rivers or undevelopable land, or collective decisions to protect land for specific uses, such as parks. Nevertheless, in general it makes little economic sense to use valuable urban land for farming, and such brutal economic logic has undermined several efforts to maintain community gardens, for example.

Urban agriculture can serve two purposes: It can generate income or food either for subsistence or for recreation. In many developing countries urban dwellers may practice farming within cities to supplement their diet, which is often meager. Crops can be grown for medicinal

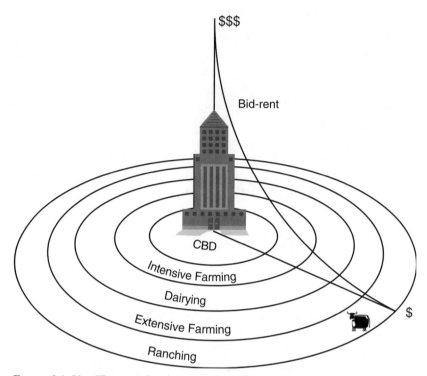

FIGURE 3.1. Von Thunen's land-use model (1826). High bid-rent values near the city center or Central Business District (CBD) make farming in cities difficult because it has to compete with other activities, like commerce, that are willing to pay more for the same space. Urban farmers usually need to squeeze into small or unused spaces within the city, such as median strips, backyards, or even walls.
Source: Figure by authors.

purposes. In Belém, Brazil, for example, some city dwellers grow crops for use in homeopathic medicine, an alternative health practice with a strong following in Brazil. A study in 2000 showed that the amount of land devoted to medicinal plants was just behind that devoted to fruits. In this tropical Amazonian city, nearly 70 percent of urban farmers are women, which is typical for Latin America (Madaleno 2000). Urban agriculture can also supplement incomes, especially when growing high-value horticulture crops such as flowers and vegetables. A study from Delhi, India, shows that urban and periurban agriculture provides the majority of vegetables in the major wholesale markets. In contrast, most of the bulky staple crops originate in rural areas (Te Lintelo et al. 2001).

Intensive horticulture can produce up to fifty kilograms of produce per square meter per year (FAO 2001). Estimates from 1993 suggest that urban agriculture adds 15 percent to the world's food supply, and this figure could reach 33 percent by 2005 (IDRC 2003).

Nowhere has urban agriculture taken on such importance as in Havana, Cuba. With the collapse of the Soviet Union in 1989 and continuation of the U.S. embargo, Cuba was forced to deal with a drastic reduction in oil, fertilizers, and pesticides (Henn 2000). Cuba had been essentially trading sugar for oil, devoting up to 30 percent of its arable land for that purpose, but that trading mechanism collapsed along with the Soviet Union. The decline in revenues meant a sharp decline in food imports, on which Cuba historically depended. Between 1985 and 1993 daily caloric consumption dropped from 2,929 to 1,863 per person. One part of the solution was to permit citizens to use, free of charge, vacant land in and around the city for farming. Urban farms now cover 12 percent of Havana's territory and provide 150 to 300 grams per capita of vegetables and herbs. Caloric intake now approaches 2,400 per person per day. Reduced fertilizer imports mean that most food is grown using organic methods, including animal manure and compost. The economic crisis in Cuba has in practice forced the use of sustainable measures and earned the city high praise for its sustainability. But if the economic climate improves, the country may revert to capital-intensive farming and return to heavy use of fertilizers and pesticides (Cruz and Medina 2003). Urban agriculture in Havana, like the victory gardens in North America and Europe, may give way to traditional rural farming if or when the crisis subsides.

Some risks are associated with urban agriculture. Use of wastewater for irrigation of urban farms can spread water-borne gastrointestinal diseases (Blumenthal et al. 2001; Keraita et al. 2003). If gardeners practice intensive methods, they may use pesticides to reduce crop loss. Contact with pesticides in densely settled areas may increase health risks, particularly if they are not handled properly (Lock and de Zeeuw 2001). Some believe, however, that the use of chemicals in urban agriculture is exaggerated and that any problems can be easily managed (Mougeot 2000). Besides chemicals, keeping livestock may increase chances of the spread of disease. Influenza, which infects 3 to 5 million people and kills approximately a quarter-million people per year, is usually passed from birds to livestock, and the most severe cases are often passed directly from avian species to human beings (World Health Organization 2003). Most influenza epidemics begin in east Asia (including the virulent

SARS outbreak), where many human beings live in close quarters with livestock. Although most tuberculosis cases are passed from person to person, human beings may also contract tuberculosis from the bovine form of the disease. If urban agriculture continues to rise, precautions must be put in place to reduce the risk of disease jumping from animals to human beings (Muchaal 2001).

In rich countries urban agriculture is usually done for recreation, but it can also serve health and sometimes economic needs. Community gardens, Schmelzkopf (1995) argues, are often initiated in times of crisis. The war gardens and later victory gardens of the world wars were meant to supplement the diets of war-torn England and, in North America, to replace food exported for the war effort. But they were also a means of lifting morale and solidifying patriotism by creating a sense of freedom through family self-sufficiency. By 1943 Americans had planted more than 20 million war gardens (Miller 2003). In the postwar boom, many victory gardens returned to grass, a remnant of wartime sacrifice.

It was not until the 1970s that many community garden efforts began in the United States explicitly to combat urban decay. Abandoned lots were converted to green spaces as a means of using unsightly properties in a fruitful way but also to build community in poor, crime-ridden neighborhoods. The growth of the environmental movement strengthened the idea of community gardens, allowing urbanites some contact with "nature" and also enabling them to grow food organically. Increasingly detached from the food-making process, the draw for community gardeners was to observe and nurture the creation of food from seeds and deliver it to their own tables. The ability to farm nutritious food for themselves, often using organic methods, continues to be an important rationale for many community gardeners (Hancock 2001; Twiss et al. 2003).

The sense of community is another benefit of participating in community garden projects. Many mothers see community gardens as safe places for children to play and as gathering places where they can meet other families. For men the community garden often acts as a gathering place for backgammon, drinking, or other activities, while gardening serves as a backdrop. Fixed investments, such as fences, ponds, tool sheds, or brick paths, tend to be minimal because most community gardens suffer from insecure land tenure. Leases are usually short-term with no guarantees that the owner, private or public, may not rescind the offer. This is especially the case in abandoned or vacant lots. Community gardens in parks are less likely to be withdrawn since the land does not have potentially "higher uses" such as housing or commercial development. An abandoned or vacant lot, if developed for housing, has much higher potential return than a community garden. One of the ironies of community gardens is that they may become victims of their own success. Because they improve living conditions in a neighborhood, they attract the interest of developers. On the Lower East Side of New York City, the community gardens of the 1970s, an era of urban decay in that area, were paved over for housing development (Schmelzkopf 1995). Insecurity of tenure remains a principal obstacle to all forms of urban agriculture worldwide (Bryld 2003).

For many supporters of community gardens, such incidents are near heresy. But within those communities, feelings can be mixed. On Manhattan's Lower East Side, for example, while many decried the loss of the green space, they also recognized the need for more and better housing (Schmelzkopf 1995). Other research has shown that "community" is not as important as the name suggests and that some users look upon community garden plots simply as a place to garden (Bouvier-Daclon and Senecal 2001).

Another limiting factor for urban agriculture is lack of available space. Community gardens, as we see, are developed on potentially valuable land, making such spaces temporary. Gardens in city parks have more secure tenure, but they may have less of an impact on improving local communities. Still, they can improve the security of parks by making sure the community has a vested interest in them. While some are large, most community gardens are squeezed into very tight spaces. In Vancouver, Canada, local community garden plots range in size from about three to thirty-three square meters and average about nine square meters (City Farmer 1997). Intensive gardening methods allow some gardens to be very productive.

Urban agriculture can occur even in apartment buildings. Rooftop gardens are one means of growing food in the city, with the added benefit of improving or slowing runoff and saving time and energy for cooling and heating (see Chapter 4 on infrastructure). Container gardening is another form of urban agriculture, what can be termed "microgardening." Gardening enthusiasts have developed ingenious methods of growing food and flowers in tiny containers. In Bangkok, Thailand, shantytown dwellers capture rainwater off roofs to feed plants growing in plastic containers. Hydroponics, or growing plants in water only, has also allowed urban farmers to produce food in small spaces. Gardeners can even use walls and other vertical spaces to grow flowers and food for sale or consumption.

In summary, urban agriculture can provide substantial amounts of food, especially high-value crops like vegetables and medicinal herbs. Farming in cities can bring many benefits beyond the food it supplies. The increased green space can cool the city, provide shelter for wildlife, and bring beauty to urban areas. Vegetation also works to improve air and water quality, and farming, if conducted properly, can also aerate compacted urban soils. Growing food near markets reduces the energy and time required to transport goods to market, which makes for fresher produce and better air quality through reduced transport. In short, urban agriculture can help to make cities more sustainable.

Yet urban agriculture contains a paradox. Although it brings many benefits to urban environments and people who live in the city, it is also a sign of crisis. The practice of farming in cities tends to be concentrated in regions of the world that are experiencing economic difficulties. Many urban farmers look upon the practice as temporary, complicated by insecure tenure of vacant properties or state-owned land. One wonders whether improved economic conditions, as in the postwar boom in North America, will mean a return to reliance on food imported from the countryside and international markets. Or it may be that the lessons learned from urban agriculture will continue to promote the idea of farming in the city, bringing with it benefits to people and the urban environments they inhabit.

4   Urban Infrastructure

*Living with the Consequences of Past Decisions
and Opportunities for the Future*

FEW URBANITES think much about the network of infrastructure
necessary to sustain city life, except when it stops functioning. A clogged
sewer, garbage strike, road under construction, power outage, or cut
telephone line are rude reminders of our dependence on infrastructure.
The engineering works that make the city livable also alter the natural
environment. Sewers and reservoirs disrupt the hydrologic cycle, impervious surfaces increase the threat of flash floods, energy use disturbs the
carbon cycle and pollutes the air. Infrastructure makes life possible in
cities, but at costs to the natural environment. Alternatives to traditional
forms of infrastructure can help reduce environmental impacts by using
better materials and smarter designs, by modifying existing technologies, or by mimicking natural systems (using so-called "green infrastructure"). Changing behavior—walking more often, for example, or
turning off lights in unoccupied rooms—can be as effective as adopt-
ing new methods and materials, or more so. As infrastructure ages in
many older cities, municipalities are working to design better, more
efficient infrastructure while maintaining a degree of equity in service.
Principles of sustainable development (the three e's of environment,
efficiency, and equity) should be applied to infrastructure development
and maintenance.

The American Heritage Dictionary defines infrastructure as "an
underlying base or foundation, especially for an organization or system." The term is applied to a variety of systems, including computer
science, government, and corporations, but is most often associated
with the physical structure that makes cities work. Roads, transit,
water and sewer lines, telecommunications, and other public works
are the infrastructure, the skeleton of services, that allows the system
to function. Historian Joel Tarr (1984) likens infrastructure to "sinews,"
the tendons of the city that bind together parts of urban areas into a
functional whole. Examples of urban infrastructure include schools,

prisons, post offices, libraries, and other public buildings. In this chapter we define urban infrastructure as the network of services (other than public buildings) that allows the circulation of people, materials, and information within the city. This includes roads, public or collective transit, water supplies, drainage, waste removal, lighting, electricity, and communications.

Public works is often used as a synonym for infrastructure. Many cities operate and maintain infrastructure under departments of public works. These services may be public, meaning that they are owned and operated by some level of government for use by the community, but some urban infrastructure is privately owned and operated, even if it is regulated by government. Most services were at one time or another private enterprises. Poor service, uncoordinated systems, and the increasing complexity and size of cities in the late nineteenth century forced many municipalities to purchase private services and make them public (Tarr 1984; Teaford 1984). Now the pendulum is swinging back. Increasingly, many public services are being privatized or replaced by public-private partnerships, with some mixing of private and public ownership, management, design, or financing. Public works are part of urban infrastructure, but so are the traditionally privately owned services, such as wireless telephone service. All services are regulated by government, in part to ensure good service but also to meet environmental obligations.

In the nineteenth century the public takeover of services happened because regulation was not effective in ensuring good service or minimizing environmental impact (especially as it affected human health). Streetcars were kept in poor condition and service was spotty, sewer lines were not planned in a systematic fashion and did not serve entire communities, and water and streams became polluted by industry and by inadequate drainage and sewer treatment. If regulation fails to maintain equity of service and minimize environmental impact, it is possible that history will repeat itself and governments will bring private services under public control. The urge to control services is strong because infrastructure is an absolute necessity for urban life. Since cities are usually the engines of national economies, the state has a strong interest in making sure they function properly to circulate capital, labor, resources, and information (Olson 1979).

Besides its economic function, infrastructure is required to make cities healthy. In densely settled areas, infrastructure makes modern cities inhabitable. Many of the services we associate with infrastructure are unnecessary in low-density rural areas and even small cities. Septic tanks work perfectly well and are in many ways more environmentally friendly (less energy- and chemical-intensive) than sewage systems and treatment plants, but they require large areas for settling, percolation, and filtration. Only with the explosive growth of cities in the late nineteenth century did infrastructure become a necessary element of urban life. In this sense, infrastructure development and urbanization are coupled processes. While rapid urban growth made infrastructure necessary, the development of infrastructure allowed for continued urban growth. One of the key components of urban infrastructure, and one of the most visible, is transit.

GETTING AROUND THE CITY: ROADS AND TRANSIT

Allowing the circulation of people and goods on common paths and roads is a critical city service. Imagine having to walk through backyards, over fences, and around various obstacles to get from one side of town to the other. We take for granted the massive investments of public and private money that permit us to travel with reasonable ease from one place in the city to another, even if those thoroughfares or buses are congested from time to time. From ancient cities down to the present, city leaders have invested considerable labor and money to assure that there are common rights-of-way in urban areas.

One of the critical infrastructure improvements in the modern city was the development of paved roads. Dirt roads remained common in nineteenth-century cities, even while other modern services, such as gas lighting, electricity, and the telegraph, were being provided. In many poor countries today, only the major thoroughfares are paved, while many of the minor streets are dirt, even if drivers are communicating on cell phones. During rainy periods, dirt roads are practically impassable. Heavy rains leave large gullies in the dirt roads that collect water and refuse.

Muddy roads are more than a messy hassle. Deluges of rain or snowmelt in the spring can render roads impassable, making the delivery of people and goods difficult and expensive. Mud mixed with horse manure, urine, and garbage is an affront to the nose but also threatens public health. Over time, several paving methods were tried to solve the problem of impassable muddy streets. Wooden paving had early success, but wooden roads would rot and needed frequent replacement. They also suffered from lack of grip. When wet, horses

FIGURE 4.1. Technology mismatch between roads and automobiles, Chicago, Illinois, 1908. As automobile ownership rose in the early twentieth century, drivers (as well as cyclists) argued for better roads to avoid getting stuck in muddy streets and for a smoother ride. The horse and wagon, on the right, is better suited than the automobile to muddy roads.
Source: *Chicago Daily News* negatives collection, Chicago Historical Society.

could, and did, slip on the planking. The wood also absorbed odors. Comments on the odor of horse urine emanating from the wooden planking during the hot summer months found their way into newspapers and frequent conversation on the streets (McShane 1988; McShane and Tarr 1997).

Cobblestone paving was in many respects an improvement over wooden planks, though the costs, especially labor costs, of installing cobblestones were high and restricted the technology usually to major thoroughfares. Like wooden planking, costs were borne by the property owners who abutted the streets. The argument that adjacent property owners would benefit most from the street paving justified the cost burden. An obvious advantage of using cobblestones is that they last considerably longer than wooden planking. Cobblestone paving put down in the eighteenth century is still in use today.

Despite the advances in paving afforded by cobblestones, the system had limitations. Like wooden planking, the stones were very slippery when wet. Spaces between them tended to collect what was dropped or washed into the street. Urban residents complained constantly about horse urine and manure. By 1900 nearly 3 million horses were in service on the streets of the United States, which required about 1.5 million stables, another serious health concern (McShane and Tarr 1997). Some municipalities instituted washing and sweeping crews to remove some of the larger waste, but most relied on the occasional rain shower to do the trick. An added concern was that cobblestones were heavy and sometimes dangerous weapons. City riots became more dangerous with this readily available arsenal. During the student riots of Paris in May 1968, cobblestones were ripped up from the streets and thrown at riot police. In the end, the *pavées* (cobblestones) themselves became symbols of resistance.

For the most part, cobblestones stayed put and the system of paving was widely adopted. In recent years some cities have benefited from the historical value of cobblestone streets. In older parts of town, the cobblestones lend a historical air, attracting tourists and increasing property values. Because the surface is not as smooth as asphalt, cars tend to travel at slower speeds, what civil engineers and urban planners call "traffic calming." Slower speeds decrease accidents, improve pedestrian and bicyclist safety, and promote roadside businesses. Although they require maintenance, cobblestones are very durable. Some have lasted hundreds of years, while asphalt roads have a usable life of about twenty years. The spaces that used to collect horse manure now percolate surface water. As cities struggle to maintain or expand aging storm sewers below impervious asphalt roads, semipervious paving systems like cobblestones or paving bricks take on new value. But costs, about double those of an asphalt road, prevent widespread use of pavers or cobblestones. With the exception of historic districts and other specialty purposes, bricks and cobblestones are rarely used.

Although visitors admire the quaintness of cobblestones, most motorists are unwilling to put up with them for more than a few hundred feet. Bicyclists fare much worse. Tires can get caught in the spaces between cobblestones, and the jarring ride can be very uncomfortable. Indeed it was the widespread popularity of the bicycle in the late nineteenth century, rather than the car, that led to demands for smooth surfaces (Armstrong and Nelles 1977). John McAdam improved roads with his macadam system, invented in the late eighteenth century,

FIGURE 4.2. Street scene in Phnom Penh, capital city of Cambodia. Most
people travel on two-wheeled vehicles, often carrying large items and multiple
family members. Increased car ownership would bring traffic to a standstill in
this rapidly growing city.
Source: Photo by authors.

which used large rocks for the substructure and small stones for the surface. The gravel surface was smoother than cobblestones and also
drained water from the roadway, but it was not as durable and was also
dusty. Later inventions used a crushed stone bed for drainage and support but surfaced the road with different materials, such as concrete, tar,
and asphalt.

Asphalt is the most common form of paving in use today. In the
United States, 96 percent of the 2.2 million miles of paved roads (another
2 million miles are unpaved) are paved with asphalt (U.S. Department
of Transportation, Federal Highway Administration 1999). This technology has been adopted throughout the world, although in some cities
only main thoroughfares use asphalt. In Phnom Penh, Cambodia, for
example, the main thoroughfares (designed as wide boulevards by the
French during colonial rule) are paved, while side streets are dirt. The
paved roads carry most of the traffic, which for the time being is primarily motorcycles and bicycles (Figure 4.2). During the rainy season,
the dirt roads become difficult to navigate. For cities like Phnom Penh,

cost is an issue. In the United States, asphalt costs vary depending on class of roadway. For new construction, the state of Florida spends $2.6 million per mile of two-lane road in rural areas and $3.5 million in urban areas. A new eight-lane interstate in an urban area costs $8.6 million per mile (Florida Department of Transportation 2005). Each year the United States spends approximately $110 billion on its highways (U.S. Department of Transportation, Federal Highway Administration 1999), the equivalent of the GDP of Ireland.

Streets and highways in urban areas have to be built to higher specifications than those in rural areas, given high traffic volumes, which increases costs and requires particular materials. Asphalt is the material of choice, but concrete paving is also used. Asphalt is a petroleum product, although 95 percent of the paving is made up of aggregate, such as gravel and stone. The asphalt industry claims that asphalt is the most recycled product in the United States. Both principal systems for paving, concrete and asphalt, have high energy inputs. A study of energy inputs for one kilometer of paved road concluded that an asphalt highway requires 7 million megajoules of power (enough to meet the daily power needs of twenty-five thousand households) and concrete 5 million (Horvath and Hendrickson 1998; U.S. Department of Energy 1995). Concrete generally has a longer life than asphalt, about twenty-five years compared to fifteen to twenty for asphalt. But if asphalt is recycled, the environmental impacts (energy inputs, ore and chemical inputs, toxic emissions and hazardous waste generation), one study shows, are less than those of concrete pavements (Horvath and Henrickson 1998).

A major drawback of paving is that it creates an impervious (or nearly impervious) surface that decreases the percolation of water through soil, thus increasing runoff and putting greater stress on storm sewers. Flooding becomes more likely since runoff occurs faster when surfaces cannot absorb water. As a result, discharge "peaks" are sharper in urban than in surrounding areas (Figure 4.3). A related problem is that paved surfaces, by not permitting percolation through soil, increase pollution runoff. Soils act as natural filters for a number of pollutants. Without the opportunity to flow through soils, water from paved surfaces runs into storm sewers, carrying a variety of pollutants and contaminants that must then be handled by treatment plants at high economic and energy costs. Leakage in the system and inadequate services means that some polluted water makes its way into water streams untreated. Hydrological models suggest that if the impervious surface of an area

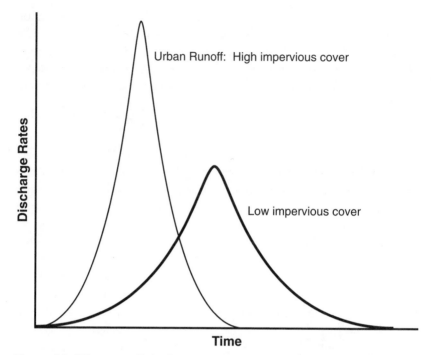

FIGURE 4.3. Water runoff discharge peaks for pervious and impervious cover. High impervious cover and storm sewers in cities tend to increase discharge peaks into local streams, increasing the chances of flooding, among other problems.
Source: Figure by authors.

gets much above 10 percent, streams become polluted and show signs of degradation, such as the loss of sensitive aquatic species. Above 25 percent impervious cover, water quality is poor, streams are eroded, stream biodiversity declines (the most pollution-tolerant species survive, while others die), and high bacteria levels restrict the use of the stream for recreation (Schueler and Claytor 1997). Excess nitrogen loading is a particularly common problem.

Human beings manufacture an enormous amount of nitrogen, primarily for agricultural fertilizers. Natural processes (including lightning, which fixes atmospheric nitrogen into nitrates) account for most nitrogen loading. Human manipulation of microbial processes through the farming of legumes (such as soybeans) also increases the amount of available nitrogen. Nitrogen fertilizer applied to farms and lawns can add significant amounts. Industrial fixation of nitrogen adds approximately

80 million metric tons per year, about 60 percent of the total global natural fixation of nitrogen (Vitousek et al. 1997).

Reducing nitrogen levels in streams requires careful application of fertilizers, but because fertilizers are nonpoint sources of pollution, this is a difficult task. Truck farms around cities often use large amounts of fertilizer to grow crops intensively on relatively expensive land. Runoff from the fields makes its way into watercourses and has deleterious effects on aquatic life and potentially on human health. Blue baby syndrome, attributed to high nitrate levels, is one consequence that can lead to death. Septic tanks, still used in many suburban areas and small towns, are a major source of nitrogen. In Wickford Harbor, Rhode Island, 80 percent of the nitrogen in groundwater supplies is estimated to come from septic tanks (Joubert and Lucht 2000).

Use and often overuse of lawn chemicals is a significant source of stream nitrogen in urban areas. Wickford Harbor officials estimate that 10 percent of the nitrogen in its groundwater comes from lawn fertilizers. Data from the Baltimore Ecosystem Study indicate that nitrogen leaching from urban and suburban areas is greater than that from forested areas of the study's watershed but less than from agricultural areas. But in suburban areas, the rate of nitrogen retention was "surprisingly high," at 75 percent of inputs (Groffman et al. 2004). Clearly, multiple factors are responsible for nitrate leaching, but overuse of lawn fertilizer, which is a documented problem, is something that most water agencies would prefer to avoid.

A common approach to reducing nitrogen loading from excess lawn fertilization is education, but some studies show that this method alone may be ineffective. A recent analysis from Columbus, Ohio, concludes that higher-income, well-educated people are more likely to use lawn chemicals than lower-income, less educated people. Even people who were aware of the environmental impact of lawn chemicals were more likely to use them than people who were not aware of the impact (Robbins et al. 2001). A study of lawn fertilizer use on Long Island also found a positive correlation between income and fertilizer use (Nassau-Suffolk Regional Planning Board 1978). The Baltimore Ecosystem Study found that subdivisions with newer homes were more likely to use high concentrations of lawn chemicals than areas with older homes. The authors suggest this could be a function of higher socioeconomic status in newer subdivisions as well as of homeowners using more fertilizer to establish lawns in newly constructed neighborhoods. An inverted U expressed the relationship between income and fertilizer application—middle-income households apply fertilizer at higher concentrations than poor and high-income households (Grove et al. 2006; Law et al. 2004). It is thus clear that we cannot rely on education alone to achieve environmental protection. Pressure to conform to community standards may have a greater effect on environmental behavior than socioeconomic standing or education. While individuals may make a difference, individual behavior must be combined with changes at a larger, collective level, either through community groups, government agencies, regulation, or legislation.

ALTERNATIVES TO ASPHALT

Some cities are experimenting with alternatives to asphalt and concrete paving, though on limited scales. Besides brick pavers, "pervious paving" offers a means of reducing runoff rates. Advances in asphalt and concrete paving have increased permeability, but other technologies may offer more effective solutions to the problem of runoff. Reinforced turf or grass pavers are one method that may work in low-traffic and low-speed situations, such as parking lots. Both use grass surfaces to increase percolation (and to cool surfaces). Grass pavers are made with a substrate of concrete, plastic, or other hard materials, often constructed in a honeycomb shape. The structure may or may not be visible on the surface, depending on the strength necessary for the parking lot. Grass is grown either within the spaces or on top of the entire structure. Reinforced turf uses a permeable support system overlaid with grass. Parts of the White House grounds in Washington, D.C., which experience high pedestrian and occasional vehicle traffic, use reinforced turf. Other facilities that need only temporary parking, such as athletic stadiums, employ reinforced turf systems. Miami's Orange Bowl, for example, uses such a system.

An added advantage is that grass pavers tend to keep parking lots cooler, an important consideration on hot summer days. Turf parking is well suited to warm climates, where temperature differences between turf lots and asphalt can be as high as 20 degrees Fahrenheit. Warm climates also permit year-round growing. An obvious disadvantage is that the lots have to be mowed, a more labor-intensive maintenance plan than filling the occasional pothole in an asphalt parking lot. Another problem is erratic parking, since drivers do not have painted guidelines for parking spots. Wet weather can also make grass surfaces slippery. An alternative is to combine traditional asphalt paving with permeable paving, allowing percolation for some if not all of the runoff from parking lots.

Treating the problem at the source, rather than using so-called end-of-pipe solutions, is often more cost effective, less energy intensive, easier to monitor, and simpler to control. Simple measures, such as painted messages on Los Angeles storm sewer outlets discouraging people from dumping in the sewers (a painted symbol next to the drain shows a fish skeleton surrounded by the message "No Dumping: Drains to Ocean"), are inexpensive compared to treating the polluted water. Some people will continue to dump motor oil in the storm sewers, but proponents argue that enough people will be discouraged that it is more than worth the minimal cost. This approach also serves to educate the public about the connection between individual behavior and environmental protection. Behavior modification on its own is rarely effective at solving an urban environmental problem. Design plays a role, too.

Increasing green material within a city is another means of increasing water absorption and decreasing runoff rates. Urban tree initiatives have been heralded for lessening multiple urban problems, from the heat-island effect to air pollution. Trees can also serve to absorb precipitation, thereby reducing peak discharges. Grass strips along roadways can also help reduce runoff. An added benefit, from a safety standpoint, is that trees tend to narrow the perspective from behind the windshield and thus calm or slow traffic. Planting trees or tearing up sidewalks for grass strips can be expensive and difficult, especially in densely built-up areas. Tree roots, if not contained, may interfere with below-street infrastructure. Trees may need to be pruned and also treated for diseases. The removal of leaf litter also costs money. Because of liabilities with falling trees and branches, some businesses are wary of using trees in private parking lots. On the other hand, property values generally increase on tree-lined streets. Research has shown that tree-lined streets may also attract more shoppers who are willing to travel longer distances and to linger longer than shopping areas without trees, which helps defray some of the costs of tree planting. Psychologists have demonstrated that urban forestry programs have even been shown to reduce stress levels and violence (Wolf 1998). In cities around the world, especially in rich countries, governments and community organizations are reassessing the value, both economic and environmental, that trees bring to urban settings.

In the most densely settled urban areas, where trees may be difficult to plant along streets, rooftop gardens offer a chance to increase green material. Such gardens reduce discharge into storm water systems by flattening out the peaks (Figure 4.3). Water is also absorbed and retained by the vegetation. Data from existing rooftop gardens show that runoff is reduced by about half, making rooftop gardens a particularly attractive proposition in cities where building codes require retention of rooftop runoff. Water quality may be improved as well, since water is filtered through soil rather than carrying pollutants and contaminants from impervious surfaces. Benefits to the building owners include reduced heating and cooling costs, since the rooftop garden acts as a thermal blanket. Evaporation in summer also cools the rooftop, putting less stress on air conditioning units. In addition, water that is not retained in the vegetation tends to be cooler in summer, which helps to reduce "thermal shock," or warm water that flows into streams, which may kill certain species (Natural Resources Defense Council 2001). Like any other green material, rooftop gardens suffer from neglect and must be regularly maintained to be effective.

A common characteristic of alternative methods of infrastructure is that they mimic natural systems to reduce environmental, and often economic, costs. In many ways, this green infrastructure philosophy is a return to nineteenth-century methods. Cesspools, for instance, were used extensively in nineteenth-century cities. The system worked because it used the natural filtering qualities of the soil to maintain water quality. Yet any student of urban history knows that nineteenth-century cities were certainly less than ideal places. Rapid growth and increased densities made such methods impossible, ineffective, and ultimately dangerous to human well-being. Municipalities responded with engineered solutions. In the case of roadways, this meant paving. Sewer pipes carried away human waste. New transportation systems and communications allowed people to live farther from one another, and from their workplaces and shopping areas, and still maintain contact. These engineering and planning principles were perfected in the twentieth century, but at a cost to the urban environment and urban taxpayer. New solutions, which may best be termed "managed" solutions, tend to blend technology and natural systems as a means of reducing environmental costs and associated economic costs.

Transportation Alternatives

While the public rarely considers the environmental consequences of underground pipes, the impact of transportation is more visible.

Smog-filled cities, especially in the heat of summer, are very real reminders that the cars we drive contribute to poor air quality (smog is produced when nitrous oxides from car emissions react with sunlight to produce ground-level ozone). Reports on global warming, tied to increasing atmospheric carbon dioxide from the burning of fossil fuels, puts transportation in the forefront of environmental issues. In wealthy countries, transportation consumes vast amounts of energy and emits large volumes of carbon dioxide and other pollutants. Approaches to fixing the problem have changed over time—from increased funding for public transportation to new urban designs that reduce dependency on the car—but most often the public and governments put their faith in new technologies.

The primary technology for cars is hardly new. Just as most lighting uses nineteenth-century technology (incandescent lightbulbs), the vast majority of vehicles use gasoline-powered internal combustion engines, first developed in France in the 1860s and refined by Karl Benz and Gottlieb Daimler in Germany in the mid-1880s. Efficiency of the engines has certainly improved, and emissions are much lower than those of the bellowing "horseless carriages" of the early 1900s, but the technology is remarkably similar. Even engine technologies thought of as new are surprisingly old. In the early days of the automobile, electric engines, such as the one developed by Thomas Edison, competed with gasoline engines as the power source for automobiles. Rudolf Diesel ran his prototypes on peanut oil, a fuel that can still be used in modified diesel engines. Henry Ford expected his Model T would run on ethanol. In the 1930s Ford built several ethanol facilities in the U.S. Midwest, convinced that alcohol from food would be competitive with petroleum. All of these attempted technologies have generated renewed interest in recent years.

Biodiesel is seen as an attractive alternative to regular diesel, especially in parts of the world where diesel fuel is scarce and expensive but vegetable oils are plentiful. Indonesia has plans for producing palm oil specifically as a biofuel. Diesel engines can run on vegetable oils with few or no modifications. In Europe many farmers use seed oil to run their tractors. Few Americans knew about biofuels until John Tickell ran a vehicle across the United States using grease from fast food restaurants. Biodiesel has many benefits. It emits no sulfur and has lower levels of particulates and other emissions then petroleum diesel does. Horsepower and torque are the same as for petroleum diesel, and mileage is the same or slightly lower (but typically higher than gasoline engines).

Biodiesel can be mixed with regular diesel in any proportion. Unlike other alternative fuels (such as hydrogen), no new fuel stations or delivery systems are required. Because the fuel is derived from plants (e.g. soybeans, palm), it is a renewable source of energy. It is readily biodegradable and nontoxic, which makes it an attractive fuel source for ecologically sensitive areas. Because the fuel can be produced from biomass, it has the potential to reduce foreign dependence on oil and the associated risks, such as oil spills, of shipping oil thousands of kilometers to market. Slightly higher costs are one reason for lack of widespread acceptance. In May 2001 biodiesel in the United States cost about thirteen to twenty-two cents more per gallon than petroleum diesel. Without the fuel tax on biodiesel, costs would be more comparable (U.S. Department of Energy 2001). The European Union has proposed eliminating biofuel taxes, as has the province of Ontario, Canada. In March 2002 the Minnesota state legislature passed a bill mandating that all diesel fuel contain at least 2 percent biodiesel by 2005. In France diesel fuel is mixed with 5 percent biodiesel. Even in small amounts, the environmental benefits can be substantial. Sponsors of the Minnesota bill noted that a 2 percent biodiesel blend would reduce particulates and sulfur dioxides by thirty thousand kilograms per year.

Alcohol has been used throughout the world to power vehicles. Brazil is the most celebrated case. Until offshore oil reserves were tapped recently, the country lacked significant petroleum resources. To fuel vehicles Brazil turned to a resource it had in plenty—sugarcane—after the oil crisis of the early 1970s sent oil prices soaring. Sugar from the cane is fermented and distilled into alcohol for fuel. By the mid-1980s more than 90 percent of new cars in Brazil had alcohol-burning engines. Fuel shortages in the mid-1990s, due in part to strikes at alcohol refineries, sent consumers scurrying back to gasoline engines. In a remarkable turnaround, by 1997 less than 1 percent of new car sales were for alcohol engines and at present less than 1 percent of the vehicles on Brazilian roads use alcohol. New offshore petroleum reserves and relatively low-world oil prices have resulted in a rapid shift toward gasoline engines. But Brazil's experience with alcohol fuel is attracting renewed interest, given its clear environmental benefits. Ethanol is nontoxic and biodegrades quickly. When mixed with gasoline, it oxygenates the gasoline, allowing for fuller combustion and a reduction in the release of hydrocarbons, a global warming agent. Although ethanol releases carbon dioxide when burned, the carbon dioxide is absorbed by plants, unlike the carbon trapped in fossil fuels. The net impact, theoretically, is no increase in atmospheric carbon dioxide, since the plants would use the carbon dioxide emissions for growth (however, fossil fuels are used to refine the ethanol). Given the reduction of carbon dioxide buildup associated with biofuels, in 2002 Germany and Brazil signed an agreement under which Brazil would produce a hundred thousand alcohol-burning engines. German car firms have agreed to pay for the incentive, a sales tax exemption of approximately $320 per car, as a credit toward reducing carbon dioxide emissions in accord with the Kyoto Protocol. The German firms will be able to count the reduction in carbon dioxide as "carbon credits." An added benefit is that the plan will stabilize sugar prices and encourage alcohol production, necessary to lure consumers back to alcohol engines. If certain sugarcane fields are dedicated to fuel production only, treated sewage can be used for fertilizer, reducing the need for agricultural fertilizers and finding an economical use for human waste (*The Economist*, 5 September 2002).

In the United States, ethanol (alcohol fuel) is used to boost octane levels of gasoline and to oxygenate gasoline to reduce carbon monoxide emissions, as mandated by Clean Air Act amendments of 1990. Annually the United States produces about 1.5 billion gallons of ethanol, mainly from corn (CNNMoney.com 2006). Ethanol production is small compared to gasoline demand, which runs between 8 and 9 billion barrels, or well over 300 million gallons a day. In some regions of the country, methyl tertiary butyl ether (MTBE), a highly toxic additive that is finding its way into the nation's groundwater sources, is used to oxygenate fuel. Ethanol is an attractive alternative to MTBE.

Biofuels are not free of problems. When combusted, they release carbon dioxide, although much of that carbon would theoretically be taken up by the plants needed to produce more biofuel, completing the carbon cycle. Agriculture can be energy intensive. Fuel for machinery and energy expended in fertilizers need to be factored into the energy costs and benefits. If crops are used for the production of fuel rather than for food, this may result in biodiversity loss, as farmers move onto land not currently farmed. The ideal solution would be to use crop wastes for biofuels, but efficiently converting material high in cellulose into alcohol is problematic.

New engine technologies are driven mainly by legislation that forces manufacturers to improve efficiency and reduce emissions. In the United States, California has been at the forefront of clean air legislation. With frequent temperature inversions (cool air trapped under a layer of warm air) and sunny days, Southern California is particularly prone to smog. The term "smog," a contraction of "smoke" and "fog," comes from Britain. These "smoke fogs" can be more than a nuisance. In 1952 a killer smog resulted in the death of four thousand people in London (see Chapter 5 for more on air quality). Smog was first recognized in Los Angeles in the 1940s, but it was incorrectly blamed on industry. In the early 1950s Dr. Arie Haagen-Smit, a Dutch-born professor of biochemistry at the California Institute of Technology, discovered that in the presence of ultraviolet radiation nitrogen oxides and hydrocarbons from motor exhaust caused smog, but automobile manufacturers initially rejected his findings. It was not until the early 1960s that car manufacturers were required to install emission controls in cars for sale in California. Since then California has instituted increasingly stringent emission requirements.

Because of the size and wealth of its market, and because California regulations influence federal policies, manufacturers pay attention to them. In 1990 the California Air Resources Board mandated that a minimum of 10 percent of vehicles for sale in the state had to be zero-emission vehicles (ZEVs). After considerable lobbying, that requirement was pushed back to 2005, although incentives in the form of increased credits were added to induce carmakers to introduce ZEVs earlier. In California owners of ZEVs are permitted to drive in carpool lanes even if they have only one occupant. Purchasers also get a variety of tax breaks and monetary incentives. Along with the ZEVs, California has mandated lower emission vehicles (LEVs) that emit 50 percent fewer hydrocarbons than a typical car. Ultra-low emission vehicles (ULEVs) and Super-ultra-low emission vehicles (SULEVs) have also been manufactured for the California market. The SULEV vehicles, which emit 90 percent less pollution than the average new car, qualify as a partial ZEV, which manufacturers can use as credit toward meeting ZEV goals (California Air Resources Board 2002; Colls 2002). Some manufacturers are now using California emissions standards in all their cars, despite the increased cost.

The drive to reduce emissions in California and in the United States as a whole is mainly to improve air quality. In Europe, by contrast, regulations to reduce vehicle emissions are motivated not only by this goal but also to reduce carbon dioxide and increase fuel economy. This broad difference in environmental priorities can be explained in part by higher fuel costs in Europe, as well as the general acceptance in Europe of the Kyoto Protocol's mandate to reduce so-called greenhouse gas emissions. In the United States, where fuel costs are a tiny fraction of household income, health concerns have been a major reason for environmental legislation linked to automobile emissions.

Hybrid vehicles are attracting attention because they can increase mileage considerably without decreasing performance or safety. By combining an electric motor with a conventional gasoline engine, cars can improve mileage by 30 percent or more. Electric motors provide the most power when a car moves from a standstill and at low engine speeds, where gasoline engines have the lowest power and produce the most emissions. While stopped, gasoline engines get the worst possible mileage: zero miles per gallon. Hybrid vehicles typically do not have the gasoline engine running at a stop and therefore conserve fuel. The advantage of combining electric with gasoline engines is that the driving range is much greater than it is with vehicles powered by electricity alone. Electric vehicles rely on power stored in heavy, expensive batteries, and the driving range is limited to about a hundred kilometers. Hybrid vehicles have much larger driving ranges, greater than conventional gasoline vehicles, and never have to be plugged in to recharge batteries. The electric motor runs off smaller battery packs that are charged by the gasoline engine when it is running and also by regenerative braking. In conventional vehicles, the power generated by braking is wasted as heat dissipated from the disc or drum brakes. In hybrid vehicles that power is captured and returned to battery packs. Most major manufacturers now make a hybrid vehicle or plan to in the very near future. The transition to hybrid vehicles is relatively simple because it requires minimal changes in behavior. Owners do not have to plug in their cars at night, as they would have to do with electric vehicles, and the hybrid vehicle's main energy source is gasoline, which is readily available through a well-established infrastructure and delivery system.

Inexpensive gasoline in North America has been a disincentive to developing technologies that improve gas mileage. The cost of gasoline in real dollars has long been in decline, although spikes in prices have caused changes in purchase and use patterns (Figure 4.4). The oil crisis of the 1970s, for example, forced oil-importing nations to reconsider their dependence on foreign sources of energy and prompted most countries to institute minimum standards for mileage. Rising fuel prices in 2004 and 2005 have, in part, driven customers away from large SUVs. In the United States, corporate average fuel economy (CAFE) standards have been unpopular with car manufacturers, especially over the past decade, as consumers have flocked to SUVs and light trucks, which are highly profitable for manufacturers. To maintain the CAFE standards,

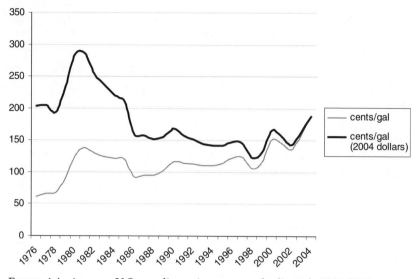

FIGURE 4.4.   Average U.S. gasoline prices, raw and adjusted, 1976–2004.
Despite steady increases in retail prices for gasoline, prices in the 1970s and
early 1980s were considerably higher when adjusted for inflation.
Source: Data derived from the United States Energy Information Agency and
the Bureau of Labor Statistics. Figure by authors.

most manufacturers have to build small, more energy-efficient cars, and
sell them at cost or even at a loss. In most other parts of the world, gasoline is considerably more expensive and efficiency is a higher priority
for consumers and governments. In Europe the average cost of gasoline
is about two and a half to three times higher than in the United States
(U.S. Department of Energy 2003b).

Diesel engines are an attractive alternative to gasoline engines since
mileage is about 30 percent greater. Many diesel engines have mileage
rates comparable to or better than hybrid vehicles, yet use an established and well-tested technology. Diesel engines generally last longer,
have fewer parts, and have stronger torque or power at the lower speeds
at which most city driving is done. Diesel fuel also requires less refine-
ment, and less energy, than gasoline. The fact that diesel engines can use
biodiesel is an added attraction. In Europe, almost half of all vehicles
sold are diesel. Diesel is widely used in heavy trucks and machinery in
the United States because of its pulling power and the longevity of
diesel engines. It is also used in tractors and farm machinery for the
same reasons. It has not been adopted in passenger vehicles to the same

extent as in Europe. Acceleration tends to be slower, the car can be nois-
ier at low speeds, and basic diesel engines can be sooty and smelly
(although manufacturers have created vehicles with turbodiesel engines
that eliminate most of these shortcomings). Diesel engines also produce
more fine particulates that have been demonstrated to cause cancer, and
they produce more smog-forming emissions than gasoline engines do.
Recent legislation to reduce the sulfur content of diesel fuel, in both
Europe and the United States, should improve emissions. Rising fuel
prices have prompted most U.S. car manufacturers to begin plans to
reintroduce diesel passenger cars into the domestic market.

Hydrogen-powered vehicles offer the hope of a vehicle with benign
emissions not dependent on limited fossil fuels. A variety of systems
have been developed, but the technology is based on producing electricity by combining hydrogen and oxygen. Fuel can be supplied as
pure hydrogen, or fuels such as gasoline, natural gas, and methanol can
be converted onboard into hydrogen. If vehicles use pure hydrogen,
the emissions are water vapor. Fuel cells offer the advantages of being
quiet with little vibration, substantially reducing moving parts and
potential for wear and tear, and increasing efficiency. The best internal
combustion engines achieve 30 percent efficiency (most of the power in
the gasoline is wasted as heat), while fuel cells have achieved efficiency
ratings of between 40 and 70 percent.

Despite the benefits of hydrogen technology, there are several factors
that limit widespread adoption of hydrogen-fuel-cell vehicles. Hydrogen
currently costs about four times as much to produce as gasoline, and a
fuel-cell-powered engine costs approximately ten times as much to build
as a conventional gasoline engine. There is at present no distribution system to deliver the volume of hydrogen necessary to fuel cars in the
world's major cities. Hydrogen, like gasoline, is volatile and requires special delivery systems to reduce the risk of explosions (U.S. Department
of Energy 2003a). It is unlikely that auto manufacturers will build a car
called the Hindenburg. Some fuel-cell vehicles are operating in government and utility fleets, and most manufacturers are currently working on
hydrogen-fuel-cell vehicles. Despite the enormous environmental benefits, few expect that such cars will be commercially viable before 2020.

Any changes made in the way Americans consume fossil fuel will
have far-reaching consequences, given that almost one-third of the
world's cars are in the United States. In 1997 Americans traveled 2.5 trillion miles in vehicles, 61 percent of those miles in urban areas. Every day
Americans travel 4 billion miles in the nation's cities. In Los Angeles

alone, vehicles are driven 238 million miles per day, more than twice the distance from the earth to the sun. For all U.S. urbanized areas, the average number of miles driven per vehicle is twenty-two (contrary to popular belief, vehicles in Los Angeles are also driven about twenty-two miles per day, the national average). For cities of more than 1 million inhabitants, Houston, Texas, ranks first, at thirty-eight vehicle miles per day. Atlanta, Georgia, is not far behind at thirty-six. For larger cities, New Orleans has the honor of being the city in which the fewest vehicles miles are driven daily, at fourteen, with New York not far above at sixteen (U.S. Department of Transportation, Federal Highway Administration 1999). Given the sheer volume of miles driven in cities, any change that reduces the use of passenger vehicles will have a broad environmental impact.

Governments and manufacturers have looked to technology to reduce energy consumption and improve air quality. While such efforts are laudatory and probably vital for future energy and transportation needs, they divert attention from factors that reduce the *demand* for transportation. Reducing demand, or conservation, is simpler and cheaper than finding technology to satisfy increased demand. Instead of spending billions of dollars on new technologies to allow increasing number of miles to be driven in private vehicles, factors that encourage people to walk, ride bicycles, or take public transportation may make such technology unnecessary or at least reduce the negative consequences of existing technologies. However, some economists make the persuasive argument that as fuel becomes scarcer, the market will find a means of transporting people in economical ways. Given the history of technology, this is probably true. But environmental groups have pointed out the costs of waiting. The burning of fossil fuels is increasing carbon dioxide emissions to the point that global temperatures are warming (a fact that the current U.S. administration has finally acknowledged, after much denial and vacillation). Poor air quality in cities is more than a nuisance—it is linked to respiratory disease and deaths. If deaths from car accidents were treated as a disease, they would be considered a frightening epidemic. Twentieth-century planning and building of infrastructure has worked to make car travel more convenient in cities, especially in suburbs, where increasing numbers of people reside.

THE MISNOMER OF LIMITED ACCESS

"Limited access highways" are something of a misnomer since they have provided people with unmatched freedom to travel by car. The term refers to the lack of cross streets, which helps to increase speeds, or at least speed limits. One of the consequences of building highways is that people can live farther away from work, travel more miles in private vehicles, and eventually cause serious highway congestion. The adage "if you build it, they will come" holds true with highways, something that highway engineers call "latent demand." Although they were originally conceived as high-speed routes in and out of the central business district, their popularity has slowed down speeds considerably, especially during rush hours. Slower speeds mean more time commuting to work, which means lost productivity. In 1997 data from sixty-seven metropolitan areas showed that Americans wasted 6.7 billion gallons of fuel and 4.3 billion hours owing to traffic congestion. The total cost to motorists was $7.2 billion. More than half of peak-hour urban interstate travel occurs under what the Department of Transportation calls congested conditions, and the problem is getting worse. From 1993 to 1997 the delay per thousand vehicle miles traveled increased from 8.3 to 9.0 hours (U.S. Department of Transportation, Federal Highway Administration 1999). The economic costs are substantial, but so are the environmental consequences. Conventional engines burn fuel even while stopped. Hybrid vehicles are superior in this respect since their engines shut off when stopped. In stop-and-go traffic, hybrid vehicles are also able to convert the frequent braking into stored energy. Also, at low speeds, internal combustion engines are inefficient and produce higher emissions per mile traveled than they do at cruising speeds.

The term "highway" comes from the "high roads" or "high streets" of England, possibly named because they were formed by mounding dirt from adjacent ditches, a means of improving drainage. These roads were the king's roads and were open to the public, unlike the private "byways." Turnpikes were toll roads, named for the spears or "pikes," forming a gate, that were turned on a central pole after the traveler paid the toll. Since it cost money to build and maintain a road, few travelers protested the idea of paying for the service. Not until the twentieth century, with the rise of the automobile (as well as bicycle use), did the idea of "free" ways take hold. Rather than collecting tolls at tollbooths, highway construction and maintenance was financed through fuel taxes. Still, toll roads have persisted, especially in smaller states that feared drivers would pass through their small territory without fueling up and hence without paying state fuel taxes.

In the past decade, tolls have become increasingly popular. Even California, the land of freeways, has instituted some toll roads. Only in the past few years has Canada built toll roads. Britain is also experimenting with toll roads. Two arguments are often made to support paying tolls: The user pays a greater share for the service, and financing is less dependent on the whim of governments. In the United States, federal money for highways is often tied to other demands (such as increasing the minimum age for alcohol consumption or the acceptable blood-alcohol level for drivers), federal control that states tend to resent. Raising gas taxes is also politically unpopular. Many environmentalists support toll roads because they reveal the cost of driving. Gas taxes are generally invisible, but having to fork over cash at a tollbooth may discourage people from driving so much. Others argue that toll roads hurt the poor and favor the wealthy. When the toll road Route 91 opened in Orange County, California, those willing to pay for the toll could avoid traffic by driving in the high occupancy toll (HOT) lanes, even if traveling alone. Locals referred to the toll road as "Lexus lanes," occupied by drivers in expensive cars. Other states have developed HOT lanes as a way of financing road construction and easing traffic congestion.

The Autobahn, begun in Germany in the 1930s, set the standard for highways in the United States and other parts of the world. By limiting the number of access points and restricting grade crossings, motor vehicles speeds could be increased substantially. The German government, including the Nazi government, argued that the economic and military benefits justified the cost. Similar arguments were made in the United States. The Federal Highway Act of 1938 provided funds for six "super highways," three running east to west and three north to south. President Eisenhower, who had seen Germany's Autobahn system, proposed the superhighways for defense purposes (American troops benefited from increased mobility on Germany's Autobahn during the invasion), and as a public works project for a country that had suffered through the Great Depression. Debate raged regarding the relative role of the federal and state governments, and whether or not the roads should charge tolls. Given that vast stretches of proposed highway would travel through sparsely settled land, the federal government concluded that tolls could not pay for the superhighway system (Rose 1979; Mertz 2002). For urban areas, the interstate system would use a hub-and-spoke system, with radial roads from city centers connecting to highways that would ring the city. Within the city, expressways would either be elevated or be dug below streets level to eliminate grade crossings. The Federal Aid Highway Act of 1944 laid out a forty-thousand-mile interstate system connecting the county's largest cities. It was not until the 1956 Federal Aid Highway Act that a suitable formula to pay for the highways was established. The federal government would cover 90 percent of the cost of construction with funds raised primarily from gasoline taxes.

In Great Britain, the Special Roads Act of 1949 allocated funds for the construction of seven hundred miles of motorways. France built a number of short auto routes in the 1950s, but because of its emphasis on rail transport did not begin to construct highways in earnest until the 1960s and 1970s. Italy, arguably the first country to build highways with the *autostrade* in the 1920s, extended its highway networks in the 1960s. The Japan Highway Public Corporation, a nonprofit government corporation, has built seven thousand miles of toll highways (also financed by fuel taxes) between major cities. National and local governments are responsible for all other roads, including the toll-based regional highways that serve as expressways for major cities (Japan Highway Public Corporation 2002). The rise in car ownership in developing countries is pressuring governments to build highways, which many are willing to do as a sign of "progress," often at the expense of more basic infrastructure such as clean water and sanitation.

Heavy government subsidy of highways has accelerated the spread of cities by making car ownership more attractive and lessening the appeal and utility of public transportation. Transportation revolutions have always remade the shape and function of cities. In North America, omnibuses in the 1830s, horse-drawn streetcars in the 1850s, and electric streetcars in the 1890s all increased speed of travel, allowing middle- and upper-class people to live farther away from the city and commute to the center for work. The maximum amount of time most people are willing to commute one way is generally an hour. In the era of the walking city, this restricted the radius of an urban area to roughly the distance people could walk to and from the center in an hour, about four miles. Horse-drawn streetcars might reach speeds of six to ten miles per hour, but with frequent stops the average speeds were closer to five to six miles per hour. Electric streetcars were considerably faster, with early streetcars reaching speeds of twenty-five miles per hour. They also had the advantage of quick acceleration but were limited by the number of stops and by a factor that limits automobile speed today: congestion. Perceptive travelers recognized that speeds were generally much faster at the periphery than in the congested core. Retailers noted the same thing and began to move out to the suburbs, particularly to intersections and major stops. No longer was it necessary to travel to the central city for all goods and services. The classic streetcar city was star-shaped, with residential and commercial development alongside radial lines extending from the core (Ward 1971; Warner 1978; Jackson 1985). Such a configuration worked well for public transportation, as it limited transfers from one line to the next and increased densities (i.e., potential riders) along transportation routes into and out of the central core. Indeed, the City of Curitiba, Brazil, one of the great urban planning success stories, has created a bus transportation system that works on the same principle of high-speed radial lines into the core (Smith and Raemaekers 1998).

One of the difficulties with rail systems is that the rails are fixed and expensive to remove to meet shifting demand. When buses were first introduced for public transportation, they were designed as "feeders" to bring people to the streetcar lines. Because of the flexibility of buses and the ability to change routes without having to tear up and reroute rails, they quickly became a favored form of transportation. One story suggests that the automobile and bus manufacturers in the United States, in concert with tire and oil companies, conspired to remove light rail from the nation's city streets in order to boost automobile and bus sales. Indeed, the case was brought to court, and in 1949 General Motors and its partner companies were convicted of conspiracy and fined $5,000. Many have argued that this was a grand conspiracy perpetrated by private business to dismantle public transit (Chomsky 1993; Snell 1974).

Others have argued that if GM and National City Lines were guilty, it was only of accelerating the inevitable (Adler 1991). By the 1920s many streetcar companies were losing money, a result of fixed fares, depreciating rolling stock and track, and declining service. Where streetcars still run, they tend to be more expensive to operate than buses. Nevertheless, North American cities such as Los Angeles, San Francisco, Portland (Oregon), Seattle, Boston, Toronto, and New Orleans have maintained or reintroduced limited streetcar lines. In New Orleans, the Charles Street trolley is a tourist draw, justifying the high maintenance costs. In Los Angeles, the streetcars have their own right of way, making light rail reasonably fast. Toronto never got rid of its streetcars, partly because the city purchased and invested large sums in its rolling stock just before most streetcar operations began to lose money and passengers to other forms of transit (Armstrong and Nelles 1986; Lemon 1985). In all cases, streetcars run on electricity and therefore improve air quality in congested parts of the city. Because the motors are electric, they shut down or work on reduced power when they are stopped. The motors are also effective for stop-and-go traffic typical of the areas in which most streetcars travel.

In most North American cities, public transit has been a difficult sell, except to the poor. In Los Angeles, the Metropolitan Transit Authority has gone to considerable expense to build a rail system, including a subway, while the more efficient buses, which mainly serve the poor, are not well maintained. The Blue Line, a light rail line that runs from downtown Los Angeles to Long Beach, covers only 11 percent of its cost with fares, while buses cover 40 percent and in some cases 90 percent with fares. The majority of riders of the Blue Line were former bus users, which cut into the goal of getting more people out of their cars (Wachs 1996). This mismatch is not atypical in the United States, as transit authorities tend to favor high-profile, expensive rail and subway projects over humble buses that go further in paying their way and serve the people who rely most on public transit.

The Los Angeles Red Line, which runs underground from downtown to Hollywood and then on to the San Fernando Valley, cost $4.5 billion to construct, which many critics have deemed too high and wasteful given the need for better bus service. Yet that is a small sum compared to the roughly $100 billion spent on highways every year in the United States. In total, public transit receives approximately $18 billion from federal, state, and local funds. Total passenger miles on public transit is 40 billion, a tiny fraction of the 2.5 *trillion* miles traveled in passenger vehicles every year (U.S. Department of Transportation, Federal Highway Administration 1999).

Despite attempts to improve speeds with radial roads from central cities to the suburbs, congestion at peak travel times has slowed traffic considerably. In response, many large-scale retailers began to move to the suburbs after World War II. Adding to growth on the periphery, increasing numbers of women in the workforce prompted employers that relied largely on female labor to relocate to the suburbs. Given the shorter distances that women were willing to commute because of their household duties (the so-called spatial entrapment thesis), employers looking to fill clerical and other female-dominated jobs followed women to the suburbs (Hanson and Johnston 1985). Currently in the United States, more trips are made between suburban locations than from suburb to city. This further erodes the utility of public transit, since densities in the suburbs are often too low to make transit viable or efficient.

The best efforts with public transit have not stemmed the tide of automobile use. Despite the billions spent on public transit infrastructure, the number of miles driven increases every year in the United States. Improved fuel efficiency has been offset by the longer distances driven. On top of more miles, vehicle fuel efficiency has been dropping over the past decade with the proliferation of light trucks and SUVs. For the time being it seems that little is moving in the direction of decreasing energy consumption for transportation.

Energy requirements for private vehicles are enormous, and any strategy to reduce the use of automobiles would have far-reaching effects on energy consumption. To illustrate, one gallon of gasoline contains 117,000 btu of energy. Assuming a car gets twenty-five miles per gallon, that is 4,680 btu or 1.1 million calories per mile. Compare this figure to the energy required to transport someone by bicycle. A 140-pound person riding a bicycle at twelve miles per hour consumes about forty calories per mile. This means that one driver consumes the same amount of energy as twenty-six thousand bicyclists. Put another way, one hour of driving consumes the same amount of energy that one person would use cycling (constantly) for nearly three years.

Increased automobile use also has an impact on human health, not only from poor air quality but from injury and death in car accidents. In the United States, more than forty thousand people die in car accidents every year. In 1990, 6.4 million people were injured in car accidents in Organisation for Economic Cooperation (OECD) countries. A study done in the northwestern United States found that "death by violence was greater from cars in the suburbs than from guns or drugs in the inner city" (Smith et al. 1998). Ironically, middle-class people were leaving the central city out of fear of crime (and race—"white flight") but as a result were more likely to die in a car accident. Increased use of cars instead of walking or biking also may mean fewer opportunities for exercise and associated health benefits (see Chapter 5).

Under certain conditions many people would prefer not to drive to every destination. But in North America, and increasingly in other parts of the world, new urban developments are designed around the car rather than pedestrians, cyclists, or public transit. Decreasing the obvious negative environmental consequences of automobile use requires not only a technological fix, such as the hydrogen car, but a redesign of urban and especially *suburban* spaces that would encourage walking, cycling, public transit, or forms of transportation other than the car. It is particularly critical in suburban areas, where populations are growing faster than in inner cities and where design and land-use patterns make cars a near necessity.

"New urbanism" is one design philosophy critical of automobile-centered planning. Its goals are simple, but the impact, if achieved, could be far-reaching. Architects Andres Duany and Elizabeth Plater-Zyberk have built several dozen towns in the United States that begin with the principle that streets should be designed with pedestrians first and cars last (NewUrbanism.org n.d.). Streets are narrow, accommodating to pedestrians, buildings are small-scale, and densities are high, qualities that urbanist Jane Jacobs recognized in the 1960s as critical to viable cities in an era of failing urban renewal projects and explosive suburban growth (Jacobs 1961). Sometimes called "neo-traditional planning," new urbanism tries to capture the qualities of older towns built prior to the automobile era. Mixed land use encourages walking rather than driving, unlike the vast suburban housing tracts built after World War II that are deliberately separated from commercial, industrial, and workplace land uses (Duany et al. 2000). Looking to the past for lessons about city living has been very fruitful for Duany and Plater-Zyberk. They have demonstrated that urban design may be more critical for reducing dependence on the automobile than hydrogen cars and the like (see Chapter 6 for more on green planning). The emphasis on building transportation infrastructure for high-speed access in and out of the city has increased dependence on the car and encouraged sprawl and high energy consumption. New, "smarter" transportation infrastructure designs are likely to receive greater attention from planners and officials in the decades ahead as one way to be more efficient with money and energy.

The circulation of goods and people within the city is an economic necessity and is usually treated as a cost. The yearning to travel, however, is also a basic human trait. The great urban planner and theorist Kevin Lynch remarked that "travel can be a positive experience; we need not consider it pure cost," since "moving through a city can in itself be an enlightenment" (Lynch 1984). Lynch reminds us that travel can be a pleasure rather than a chore, but only if we "pay attention to the human experience: the visual sequences, the opportunities to learn or to meet other people." Being enclosed in a steel cage alone among strangers on a congested freeway would not, by Lynch's definition, be a pleasure. Humanizing the travel experience within cities remains a difficult challenge that goes beyond transportation technology or infrastructure and addresses larger questions about what makes a city enjoyable.

WATERING AND DRAINING THE CITY

The hydrologic cycle ensures a constant flushing of water through urban systems, a process of enormous value given what cities do to the quality of water. Cities rely on the hydrologic cycle to remove polluted water and replace it with a clean supply for everything from drinking and bathing to industrial production. In this sense, urban areas are using the power of the sun, the driver of the hydrologic cycle, to cleanse themselves and replenish water supplies. Controlling the flow of water into and out of the city has long been a preoccupation of city leaders. Engineering infrastructure has allowed urban dwellers to capture and remove water as needs arise.

Since ancient times, cities have had to look far afield or underground for clean water. Nearby streams quickly become polluted by dense human activity. Wells were and are the simplest way for many city dwellers to get clean, safe water. The filtering properties of soil remove many of the impurities found in surface water, though groundwater may also pick up minerals that harden water and may include toxic chemicals such as arsenic. For most of human history, open wells have worked fine in cities if supply was sufficient. Overdrawing of well water has forced many cities to look for other supplies. The Roman aqueducts, an engineering marvel, are a testament to the great expense cities have shouldered in order to ensure an ample and ready supply of fresh, clean water. Southern California draws on water from hundreds of miles away in the Owens Valley and Sierra Nevada Mountains. New York City is spending $6 billion on a sixty-mile tunnel (about twice the length of the channel tunnel or "Chunnel" between France and England) to bring 1.3 billion gallons of water daily from reservoirs in upstate New York (Gandy 2002).

Too often cities have treated underground reservoirs as inexhaustible, and the result has been overdrawing and waste. Throughout the United States, municipalities are faced with depleted groundwater supplies. Coastal cities are particularly at risk because their groundwater reservoirs may fill in with adjacent salt water (termed "saltwater intrusion"), making the reservoirs useless even if left to regenerate. For more than a century along the Atlantic coast of the United States, people have observed increased salinity of groundwater from saltwater intrusion. Continued growth is expected to increase the potential for saltwater intrusion in groundwater supplies that support 30 million people at present (Barlow and Wild 2002).

The earth is a watery planet, but less than .003 percent of it is readily available as fresh water. Dependent on huge sums of fresh water, cities, especially in coastal areas, are beginning to eye the oceans as a water source. Technology to convert saltwater to fresh has been around for a long time, but the vast energy required to desalinize ocean water has restricted the practice to areas where energy is cheap and water is scarce—primarily the Middle East. In Saudi Arabia, twenty-seven desalination plants produce 715 million gallons per day, about 70 percent of the nation's drinking water, most of it destined for urban areas (Saudi Arabia Information Resource n.d.).

Other cities that face scarcity are now considering desalination. Tampa Bay, Florida, recently opened a desalination plant that will provide water to about two hundred thousand homes and supply 10 percent of all households. By sending saltwater through a series of membranes at high pressure, organics and salt are removed. The cost is about twice that of groundwater, adding about $7 on average to a household's monthly water bill. Although it puts less pressure on groundwater supplies, concerns remain about the impact of the leftover concentrate, higher in salt and organics than ocean water, on the local ecology (Davis 2003). Santa Barbara, California, and Key West, Florida, have smaller systems, but they have been used only on a limited basis and are in place mainly for emergency or severe drought conditions. Improved technology and lower prices for filtration of saltwater will probably make the Tampa Bay project an attractive proposition for other coastal cities.

Faith in engineering has led to technological solutions to increase supply. Over the course of the twentieth century most industrialized countries increased per capita water consumption. Agriculture accounts for the largest use of fresh water, currently about 65 percent. Industry and domestic use account for the remainder. Conservation of water in agriculture would free up more sources of fresh water for industry and domestic use in cities, if the sources were located within a reasonable and cost-effective distance of cities. Encroachment into semiarid regions has been taking place for more than a hundred years in the United States and for millennia in other parts of the world. Abundant sunshine, long growing seasons, and cheap, often subsidized water for irrigation have proved to be a very profitable combination. Cities that have grown up in semiarid regions, such as the American Southwest, have to compete with agriculture for water. The solution has been to dig deeper wells or build longer aqueducts to water supplies. Los Angeles's famed tapping of the water in the Owens Valley, fictionalized in the movie *Chinatown,* spelled ecological disaster for the valley and its residents (Reisner 1993). Water efficiency has been an increasing concern of water providers, given the simple notion that a liter of water conserved is the same as, and usually better than, a liter of water provided. Low-flush toilets, now standard in most building codes, significantly reduce water consumption in homes, since the toilet typically uses more water than any other appliance. One community group, Mothers of East Los Angeles, has promoted the use of low-flush toilets because of their multiple benefits to residents. In a program sponsored by the Metropolitan Water Board of Los Angeles, residents can bring in their old toilets in exchange for free low-flush models. The new toilets reduce residents' water bill and save the city from having to build expensive projects to increase supply to the growing Southern California region. The Mothers of Easter Los Angeles receive a payment for each old toilet, and the broken porcelain is used in road construction. The whole scheme also provides jobs (Mothers of East L.A., Santa Isabel n.d.). The program is a classic example of what sustainable development is intended to achieve—reduced consumption of resources, money savings, and environmental benefit.

REMOVING WATER

The modern industrial city has necessitated engineered solutions for surface water removal. The regrading of surfaces, removal of vegetation, filling in of wetlands, logging of watersheds, and creation of impervious surfaces have all worked in combination to increase the potential for flooding in cities. In some cases severe floods have caused deaths as people were swept away in rushing waters or killed by debris. More often the deaths were less dramatic and sudden, brought on by dirty water and associated diseases. The typhoid epidemic of Chicago in 1885, for example, followed a major flood in the city. Water from the Chicago River mixed with contents of privy vaults and cesspits, a potent mixture that provided a perfect breeding ground for the disease vector (bacteria). Chicago's flat topography, and the fact that it was a former bog on the western shores of Lake Michigan, did not help matters, since it took a great deal of time for standing water to drain. While the death of citizens and public health of cities concerned city leaders, more often they were motivated by the enormous property damage that could result. Hundreds of millions of dollars have been lost to flood damage in major cities, especially since many industries and warehouses are located close to bodies of water so as to take advantage of their benefits.

Given the high value of waterfront property in the industrial age, the idea of moving property off the floodplain has rarely been considered a good solution. Typically the answer has been to build a floodwall at a height above an extreme event, put some pumps behind the walls, and hope for the best. Montreal, for example, built such a system of levees and pumps (at public expense, primarily to protect private companies) after a flood in 1886 inundated an industrial part of the city. After considerable study, engineers demonstrated that the floods were caused partly by ice jams in the St. Lawrence River but also by extensive logging in the watershed. But the only thinkable solution, as has been the case in other cities, was to engineer a way out of the problem by building a levee, rather than replanting the watershed or moving valuable property off the floodplain (Boone 1996).

Cities make their own floods. Impervious surfaces decrease percolation and increase runoff rates, resulting in greater discharge peaks. In the United States, the Army Corps of Engineers has been responsible for flood control of most rivers, as dictated by the Flood Control Act of 1936. In the twentieth century the federal government has provided funds for flood control in urban areas based on the need to protect property. One of the ironies of flood control, as geographer Gilbert White pointed out decades ago, is that it encourages further construction on the floodplain, thereby increasing the potential for damage and injury (White 1958). Indeed, the work of White and others has encouraged the Corps of Engineers to look for other solutions to flood control, including better floodplain management and natural checks on floods such as reforestation.

Some municipalities have addressed the flooding issue by exploiting natural processes to reduce peak discharges, or rapid runoff that results from alterations of the natural hydrology. Catchment basins, for example, retain water after heavy precipitation and allow the water to percolate through the soil. This decreases runoff but also works to recharge groundwater. By allowing the water to move through soil, a great natural filter, it is purified of many common surface water pollutants.

New methods mimic natural processes to improve water supply and reduce flooding. Wetland restoration and re-creating urban rivers are two approaches that can mitigate flooding and improve overall ecosystem health. Wetlands are great natural "sponges" that slow the percolation of water into streams and rivers, thereby reducing the rapid runoff that can lead to floods. They also serve to remove pollutants from water. Environmental engineers are studying a number of different plants, many of them native to wetlands, for their ability to remove toxic chemicals from the ecosystem, a process called phytoremediation. By absorbing toxins in their roots, the plants remove contaminants from soil, water, or wetlands and can then be disposed in an appropriate manner, such as a hazardous waste landfill (Tesar, Reichenauer, and Sessitsch 2002). Phytoremediation can be a very cost-effective solution for removing contaminants from the environment, another clear benefit of plants and biological processes to human welfare.

While many people regard this shift to using natural systems for ecosystem benefit as novel, the ideas are not new. Frederick Law Olmsted's design of Boston's "emerald necklace" of parks, for example, included fens that reduce flooding and filter wastewater (Spirn 1984). The Olmsted brothers' plan for Baltimore (never fully implemented) also recognized the importance of vegetated riparian buffers for slowing discharge and improving water quality (Olmsted Brothers 1987). What killed those ideas in favor of engineered solutions like flood control was the enormous economic value of urban land. Private and public forces worked to convert "unimproved land" to "higher uses," relying on technology to mitigate any "natural hazards" that might result from those changes. Urban real estate was simply deemed too valuable to be set aside for riparian buffers. Maturing industrial societies, grown used to the feats of technology and human ingenuity, believed that engineering could solve any problem imposed by nature. But as many hazards researchers have shown, a natural hazard is not a "hazard" unless humans consider it one. Floods perform important ecosystem services, such as the depositing fertile silt on floodplains, but if they damage human life and property, they become natural hazards (Palm 1990). Late twentieth-century thinking has begun to question the concrete channeling of rivers and filling in of wetlands, but the legacy of earlier decisions leaves cities with engineering works built to alter natural systems solely in the interests of human welfare (Gumprecht 1999). Storm and sanitary sewers are one example of that legacy.

FROM DUNG HEAPS TO CONCRETE CONDUITS: DEALING WITH HUMAN WASTE

In medieval times, Paris was ringed with dung. The suburbs of that era, and even those of today, were put to less glorious uses and inhabited by humbler classes than the celebrated center. Human waste from the homes of central Paris was simply dumped at the periphery, and over time the dung ring attained considerable height. For fear it would be used for gun emplacements by an enemy force, the ring of dung heaps was incorporated into the city's fortification during Louis XIII's reign (Reid 1991). "Nightsoil," so named because the privies were usually cleared of their contents at night, was also carted out of the city and used for agriculture, and it was used within the city, along with urine, for a variety of industrial purposes (Guillerme 1988). With the rapid growth of cities in the nineteenth century, these methods for the removal of human waste became inadequate. In 1801 more than half a million people lived in Paris, a figure that doubled by 1846. More important, the population density of Paris increased to thirty-one thousand per square kilometer, greater than present-day concentrations. The major dump at Paris, the infamous Montfaucon (which accepted all types of waste, including the bodies of criminals hanged on nearby gallows), was within three hundred meters of the city by the mid-nineteenth century, and the stench and liquids that emanated from the dump forced Parisians to look for another solution. The cholera epidemic of 1832 also prompted the city to search for ways to safely remove waste. Underground sewers were the answer, a plan that was initiated on a limited scale in the early nineteenth century and expanded during Haussmann's famous "grand works," a period of substantial redesign, including the grand boulevards that now characterize Paris (Barles 2002; Ratcliffe 1990).

Sanitary sewers differ from storm water sewers in that they carry household and industrial waste. In other words, they work to keep cities sanitary, a reference to the public health principles that drove early efforts to create comprehensive sewer works. Paying for and constructing underground pipes to remove human waste was a long, hard-fought battle. In Paris, Baron Haussmann originally resisted the idea of human feces in his grand sewers. On his side were the engineers, who argued that human waste could be used as an effective fertilizer for farms that grew food to feed the city, what has been termed "mutalism" (Barles 2002). This idea had longevity: the last sewer farm outside Paris was not closed until 1999.

In Paris and other cities, a second factor that slowed the construction of sanitary sewers was the vested interest of nightsoil collectors. These were usually private companies that collected fees from households to remove waste from privies and then sold the waste to farmers for fertilizer, valorizing a service and a good at the same time. Nightsoil collectors operated in all major industrial cities, and their fear of losing their profits gave them reason to resist underground sewers. What tipped the scales in favor of underground sanitary sewers was disease (although in some cities, like New Orleans, storm sewers were also important for removing standing water, which could be a vector for disease) (Colten 2005; Colton 2002). Cholera epidemics in the 1830s, continuing bouts of typhoid, and yellow fever epidemics combined with new thinking on sanitation forced cities to examine their wastewater practices. In France and Britain new theories in the 1830s and 1840s discounted fatalism, God's will, and poverty as the root causes of disease. Unsanitary conditions, including filth, bad smells, and putrefying matter (miasmas) were acknowledged as the principal causes of disease, and so the solution was deemed a matter of improving sanitation. By the 1850s the sanitary idea was well entrenched on both sides of the Atlantic. Municipalities passed a series of ordinances to prohibit dumping of garbage and human waste in the streets and public areas. Edwin Chadwick, a particularly influential proponent, argued that public works, including underground sewers, were critical for sanitation. These ideas resonated in the municipal councils of Europe and North America, prompting many cities to investigate sewer technologies in the 1850s and 1860s (Melosi 2000; Waring 1883).

In 1849 the Metropolitan Sewers Commission was established in London to build sewers, although the plans were for incremental construction. In 1859 the Metropolitan Board of Works began a more comprehensive sewage system for London. In France the monumental sewer works, begun in 1850 under the watchful eye of Baron Haussmann, became a model for the world. In the United States, Brooklyn began the first planned underground sewer system, drawing on the sanitary ideas put forward by Chadwick and others. The Brooklyn sewers were followed by sewers in Chicago (1859), Providence (1869), Cincinnati (1870), Indianapolis (1870), New Haven (1872), and Boston (1876) (Melosi 2000). By the 1880s most large U.S. cities had sanitary sewers, Baltimore being a notable exception (Boone 2002).

Early adoption of sanitary sewers saved lives by improving public health, but most of the older systems used combined sewers where both storm water and household waste flowed through the same pipe. Although Baltimore did not decide to build sewers until 1905, it used the latest technology and built a separate system, one for household waste and one for storm water. The problem with combined systems is that storm water and sewage are sent to the treatment facility, adding unnecessary cost. In the United States, until the late 1980s, storm water from municipal systems could be dumped into water bodies untreated.

Since the 1987 amendment to the Clean Water Act, larger municipalities with separate sewer systems require permits to discharge storm water into water bodies. The EPA grants permits only when plans are in place to reduce pollution from storm water (EPA 1996). Nevertheless, standards for the treatment or prevention of pollution are not as stringent as those for sewage.

Early sanitary sewers dumped untreated sewage into water bodies in the hope that dilution would remove any harmful effects. By the 1880s, however, the miasmatic theory of disease was in rapid decline, and most physicians began to accept the bacteriological origins of disease (Melosi 2000). Recognizing that disease could be spread through contaminated water, public health officials and engineers began to look for ways to treat sewage before releasing it into water.

Since most cities had nightsoil service at one time, many municipalities dabbled with the idea of sewage farms, where wastewater would be used to irrigate and fertilize farm fields. In Sydney, Australia, a sewage farm was created in the late 1880s on government land that had formerly been used for dumping nightsoil. Initially the farm was very successful, but a sixfold increase in sewage runoff by the early twentieth century caused frequent flooding, making the land impossible to farm. Complaints from local residents about the smell, along with the high volume of sewage, forced the government to close the facility in the 1910s (Beder 1993). In 1896 Berlin modeled its sewer system on that of Paris, using sewage farms. Pasadena, California, ran a very successful three-hundred-acre sewage farm. It proudly sold burlap bags of dried sewage (called "poudrette") until objections to the smell forced the city in 1914 to stop sending sewage to the farm and direct it instead into the Pacific Ocean. Most sewage farms in North America were closed in the early twentieth century, but Boston, bucking the trend, recently opened a plant to convert sewage sludge into fertilizer (Massachusetts Water Resources Authority 2003).

Coastal cities around the world have long relied on the oceans for dilution of waste. Even when released into the oceans, fecal matter can return to shore and pollute beaches, forcing public health agencies to close them, especially in the summer months. Sewage treatment plants can also overflow, especially during heavy rainfall and with combined systems, with the result that untreated sewage makes its way into water bodies. Aging infrastructure increases the chances for leakage and overflow. Often it is not until they reach a state of crisis that municipalities begin to address sewer problems (Davis 1992). As part of the invisible city, sewers tend not to receive the attention of politicians or citizens as much as infrastructure that sits visible aboveground.

At a minimum, sewage treatment removes solids, a process called primary treatment. Historically, this was achieved through sand filters, a process that had been used in early water purification schemes. In some cases, such as in Sydney, the land was also farmed to take advantage of the fertilizer. Present-day methods pipe the sewage to holding tanks where the solids settle, forming sludge. The sludge may be disposed of in a landfill or is sometimes scattered on public land. Because primary treatment does not remove most of the organic matter, the effluent may be sent to oxidation tanks, or secondary treatment, to remove the majority of organic matter. The effluent is usually treated with chlorine to kill any bacteria and then released into a water body downstream from water intakes. Secondary treatment does not remove other harmful contaminants, such as nitrogen or phosphorous, which can lead to eutrophication. Tertiary or advanced sewage treatment methods can be employed to remove minerals, but they are costly to build and operate. Wealthy, densely settled European countries are more likely to use tertiary treatment and have a greater percentage of their population hooked up to sewage treatment facilities. In Denmark, 99 percent of the population has connections to sewage treatment plants, while in Canada the figure is approximately 66 percent. In sub-Saharan Africa, by contrast, only 2 percent of cities have sewage treatment, and of those only 30 percent of the plants are operating properly (United Nations Environment Programme 2000).

The high cost of construction and operation makes it difficult for poor countries to provide sewage treatment, threatening the health of city dwellers who live near streams and other outfalls. Many shantytowns, where the poorest residents live, are built in riverbeds where the threat of floods discourages conventional housing. Rivers act as both sewers and water supply, making these populations particularly vulnerable. Even if money is available, building infrastructure in shantytowns is difficult because housing is built without accounting for basic services. Building underground sewers in shantytowns is difficult and expensive unless the housing is removed. Regardless of the technical difficulties, poor communities, especially illegal shantytowns, are unlikely to receive municipal services. Shantytown dwellers have little political clout. Municipalities are reluctant to provide services, since it would legitimate what are in many cases illegal settlements. In some cases, however, limited services are supplied to shantytown dwellers in exchange for votes, creating a system of dependency that rarely fulfills the need for basic services (Thornton 2005).

While municipalities struggle to keep up with population and physical growth of cities, some cities in western Europe have to manage an entirely different problem—shrinking cities. In Leipzig, Germany, population is declining while the underground infrastructure remains the same. Since the water supply and sewers were designed for a certain "load," or level of service, the decline in demand makes the existing infrastructure inefficient at best and difficult to operate at worst. After the reunification of Germany in 1989, East Germans flooded into West Germany, leaving more than a million empty apartments (Großman 2004). Leipzig alone lost a hundred thousand people after reunification, some to West Germany and, recently, even more to suburbanization (Shrinking Cities n.d.). Germany is not alone in this. Shrinking cities are found throughout the world. In the United States, Detroit suffers from problems similar to Leipzig's, with a declining central city population that reduces the tax base and makes the maintenance of infrastructure difficult. While Pittsburgh has a more vibrant economy than Detroit, it has had to adjust to a loss of nearly half of its residents since 1950 (Tarr 2002). Deindustrialization, particularly the decline of steel manufacturing, played a large part in the shrinking of Pittsburgh, although the city has a long history of residential and industrial suburbanization (Muller 2001). Building new infrastructure on the fringes while existing infrastructure in central cities is underused is an inefficient use of resources.

Given the rise and decline of cities and the high cost of sewage treatment, engineers and environmental planners have looked for simple, flexible, and cost-effective alternatives. Ironically, these methods are more likely to be used in affluent parts of the world because of the ecological benefits they provide, but they have been used in poor countries as well. One example is the use of artificial wetlands. Wetlands are great natural filters, removing wastes and sediment from water bodies. They are also susceptible to draining, since historically people have perceived wetlands as "wastelands" and even as breeders of disease. People also drained wetlands because the soil, rich in organic matter, can be very fertile (see Chapter 6). Some cities are now experimenting with artificial or constructed wetlands to filter sewage naturally. In Arcata, California, an artificial wetland was constructed to treat about 2 million gallons of effluent a day without the use of chemical inputs and with negligible amounts of human-generated energy. The system is economically and ecologically efficient, a good example of sustainable development at work. Arcata has some of the lowest sewage bills in the country. The artificial marshes are home to bird sanctuaries and also a town park. Phoenix, Arizona, in order to meet new Clean Water Act requirements, decided to build an artificial wetland at an estimated cost of $80 million rather than pay an estimated $625 million to upgrade an existing sewage treatment facility. In a smaller pilot project, forty-five species of birds were observed at the wetland (Gelt 1997). About 150 towns and cities in the United States now use artificial and natural wetlands to treat wastewater.

Cities alter the ecosystem drastically, and in a sense it is absurd to think that natural systems alone can ensure the health of such densely settled areas. In the suburbs, the situation is different. On large lots with low residential densities, septic tanks are a possibility. Some have argued that the septic system actually *encouraged* the development of low-density, far-flung suburbs, because the technology freed the housing developments from the ties and regulations of a sewer system usually controlled by central city administrations. In the United States, especially during the suburban housing boom after World War II, poor design and lack of maintenance led to frequent failure of the septic systems, resulting in polluted streams, eutrophication of lakes, and in some cases infectious diseases such as hepatitis. Because many of the homes were insured by the federal government through the Federal Housing Administration and Veterans Administration, the federal government was forced to spend billions in subsidies for the construction of sewers in low-density residential areas where efficiency is low and price is high (Rome 2001). Yet the high cost of building suburban infrastructure has not been an effective deterrent to urban sprawl (we return to this issue in Chapter 6 when we examine the impact of green governance on reducing the cost of infrastructure).

Infrastructure is taken as a given in the cities of rich countries. It often goes unnoticed because it is underground or simply because it has come to be taken for granted. The economic and environmental costs of "hard infrastructure" have prompted alternatives, including the use of "green infrastructure" or natural systems to clean our water and air, absorb our wastes, reduce flooding, and decrease automobile use. In the twentieth century planners and engineers relied increasingly on technological fixes to problems associated with high-density living. While improved technology can bring real benefits to the quality of life and health of citizens, new approaches to infrastructure demonstrate that it may be possible to provide the same services using managed "natural"

approaches as engineered solutions, but with fewer economic and ecological costs. Providing basic needs in cities of the developing world will
continue to be a difficult challenge, given the lack of planning, rapid
growth, limited funds, and consistent poverty that plague poor countries. Examples from certain visionary administrations, such as Brazil's
Curitiba, give us hope that intelligent decisions can provide simple and
effective means of providing essential services and infrastructure to
people living in rapidly growing cities in developing countries. Rather
than exporting expensive infrastructure solutions to poor countries,
wealthy nations could learn from the resourceful solutions originating
in the cities of the developing world.

5   Healthy Cities and Environmental Justice

STRONG ANTIURBANIST sentiment, especially in North America, is tied in part to the belief that cities are polluted, unhealthy places to live. The clean air and bright blue skies of the countryside are one reason people flee the city, if only as far as the suburbs (Kaplan 2001; Rybczynski 1995). Even though that flight, usually by automobile, contributes to the smog that blankets large cities, many people justify it on the grounds that the suburbs are a healthier, cleaner place to live. This pattern of enjoying environmental and heath benefits without paying the environmental and health costs remains a point of tension between city and suburb. Inner-city dwellers understandably do not want to be the sink for metropolitan waste.

The problem of disproportionate benefits and costs is confronted by the environmental justice movement, which does not necessarily strive to make suburbs less healthy but seeks to make life better for everyone. More immediately, the environmental justice movement works to improve quality of life for people bearing a disproportionate burden of environmental costs. In most environmental justice struggles, the issue of health is paramount. Fear of health problems emanating from a toxic site or polluted air tends to mobilize communities under the banner of environmental justice (Bullard and Johnson 2000). One of the ironies of the environmental justice movement is that it draws attention to the environmental risks of inner cities, providing another reason for people to leave the city behind for the perceived safety of the suburbs. But are cities unhealthy places to live?

ARE CITIES UNHEALTHY?

In general, cities are considered unhealthy places, but the reality is more complicated than that. Respiratory problems such as asthma have higher incidence rates in urban than in rural areas because of poor air quality in cities, but some diseases are less prevalent. Malaria, for

TABLE 5.1. Urban and Rural Infant Mortality Rates in Selected
Developing Countries

Country	Infant Mortality Rate	
	Urban	Rural
Peru	29	60
South Africa	33	52
Brazil	42	65
Turkey	43	59
India	49	80
Cambodia	72	96
Benin	73	105
Rwanda	78	124
Ivory Coast	85	124
Mozambique	101	160
Mali	106	132
All developing countries	62	83
England[1]	5.6	4.6

[1]Included for comparison of infant mortality in a wealthy country.
Source: Demographic and Health Surveys (2004).

example, tends to decrease with urbanization, as built-up areas provide
a less agreeable breeding environment for the mosquitoes that carry the
disease. Better access to healthcare in cities means that some things,
such as the death of women during childbirth, become less frequent in
cities than in remote rural settings. Infant mortality in most poor countries is lower in cities than in rural areas (Table 5.1). In Nepal, for example, the infant mortality rate (the number of deaths of children under
age one per thousand live births) for urban areas is sixty, while in rural
areas it is one hundred (World Health Organization 2004b). A study in
Andhra Pradesh, India, shows an infant mortality rate of forty-seven
in urban areas and seventy-nine in rural areas, while child mortality
(the number of deaths of children under age five per thousand children
under age five) is eleven in urban areas and twenty-six in rural
(Nagdeve and Bharati 2003). Child mortality in Niger shows a similar
disparity, with a rate of 293 in rural areas compared with 168 in cities
(World Health Organization 2003b). Unclean water, poor sanitation,
and inadequate access to healthcare are some of the factors that contribute to higher rates in rural areas.

Infant mortality is often used as a human development indicator because it reflects general characteristics of a population's health. It also serves as a proxy for environmental conditions, since infants are more susceptible than adults to disease. For these reasons researchers use infant mortality rates to investigate urban environmental conditions when other types of data are not available. In the past, infant mortality rates tended to be higher in cities than in the countryside. Evidence from Great Britain shows that not until the 1920s did infant mortality rates in cities drop below rural rates (Woods 2000). In Ontario, Canada, infant mortality rates remained above rural rates until 1920 (Haines and Steckel 2000; McInnis 1990). The squalid, crowded conditions of industrial cities, coupled with poor sanitation, were the principal causes of the high urban rates (Mercier 2004; Mercier and Boone 2002).

A long history of disease in crowded cities may be one reason why the association of cities with poor health endures. Ancient Rome suffered from cholera, typhoid, typhus, and malaria, even though the Romans built the Cloaca Maxima, the grand sewer, which is still in use today, in the sixth century B.C. Although the ancient sewer illustrates Roman foresight, the lack of hookups above first floors and to tenements meant that disease could and did spread easily (Mumford 1961). Medieval cities, while perhaps not as infested with vermin as the popular media would have us believe, were nevertheless centers of disease—including the Black Death (bubonic plague) in the fourteenth century—associated with crowded settlements and inadequate sanitation. As we saw in Chapter 4, the sanitation methods of early cities were very crude, relying on the collection and disposal of human waste and garbage beyond city walls. In Paris the horrific suburb of Montfaucon accepted all types of waste, including the bodies of criminals collected from the gallows (Reid 1991). In London, people began to move to the suburbs in the seventeenth century after the plague and the fires of 1665 and 1666. Some Parisians, too, moved to the "purer air" of the Bois de Boulogne and Parc Monceau in the seventeenth century (Jackson 1985). The flight from city to suburb to escape disagreeable living conditions has a long history.

In North America, nineteenth-century cities were as unhealthy as earlier cities in Europe, if not more so. San Francisco suffered from smallpox, tuberculosis, syphilis, and even bubonic plague well into the twentieth century. Crowded Chinese neighborhoods, hemmed in by racism, were particularly affected by epidemic diseases. For the city as a whole, infant mortality in 1869 stood at two hundred per thousand live births (Craddock 2000). An analysis of 1880 data shows an even higher rate in Baltimore, Maryland. In this port city, one in four infants died before reaching its first birthday. Poor sanitation was a major factor. Infant deaths were clustered in the low-lying areas of the city, and since Baltimore did not begin building a comprehensive sewer system until 1905, water supplies were easily polluted with human and animal feces and other household wastes (Boone 2002; Hinman 2002). Cholera, typhoid, and other diarrheal diseases were particularly deadly to infants, the vast majority of whom died from gastrointestinal infections. Parents recognized the coincidence of infant death and living in the lowlands, and those with the means began in the 1860s to move to the higher piedmont to improve the life chances of their children (Olson 1997). The British hill stations in nineteenth-century India served a similar purpose, as colonial administrators tried, not entirely successfully, to escape the disease (as well as the heat and the "natives") of coastal cities (Kenny 1995; Kenny 1997).

Poor sanitation was only part of the reason for the high incidence of death. Research in Montreal and Ottawa, Canada, shows that infants born to French Canadian families were more likely to die before their first birthday than infants in English-speaking homes, after controlling for socioeconomic status (Mercier and Boone 2002; Thornton and Olson 1991; Thornton and Olson 2001). Clearly it was not the French language that was dangerous to infants' health, but the cultural practices of French-Canadian homes, such as early weaning, may have increased risks for infants. These studies demonstrate that the external causes of infant mortality, and of disease and death in general, can be agonizingly difficult to pinpoint. The figures from nineteenth-century cities show that they were more deadly than rural areas, but the reasons for those deaths were not always clear.

Infant mortality rates in poor countries today are similar to rates in wealthy countries in the past. Birth spacing, family size, nutrition, breast-feeding practices, access to health services for mothers and children, socioeconomic status, and environmental conditions are all significant variables for infant mortality rates (Rutstein 2000). The "urban effect" of lower infant mortality rates, in other words, is tied to multiple variables. Within cities there are variations that reflect differences in living conditions, socioeconomic status, and behavior. In Porto Alegre, Brazil, for example, infant mortality is three times higher in the *favelas* or shantytowns than in the established urban core, reflecting both the poverty of the population as well as poor housing and inadequate urban services (United Nations Population Fund 1996). In the United States, infant mortality rates are lowest on the fringe of large metropolitan areas. For these suburbs, the rate is 6.1, while in the central cores of large urban areas in the Northeast and Midwest, rates are above 7.5. For nonmetropolitan counties, or rural areas, the rates range from 7.5 to 7.7, or 25 percent higher than for suburbs of large cities (Centers for Disease Control and Prevention, and National Center for Health Statistics 2001). In the case of wealthy countries, living conditions may play a role in differing infant mortality rates, but other factors, such as access to emergency care or socioeconomic status, can be more critical determining factors.

History shows that high population densities in cities can lead to the quick spread of diseases that sometimes lead to epidemics. Contagious diseases such as influenza spread quickly in crowded conditions, where human beings come into close and frequent contact. Tuberculosis, another major killer, tends to be transmitted more successfully in urban areas than in rural. Sexually transmitted diseases are more prevalent in cities than in rural areas because urban dwellers tend to be sexually active at a younger age and have more partners than their rural counterparts (United Nations Population Fund 1996).

The evidence suggests that in some respects life in cities is healthier and in others less healthy than life in the countryside. A simple rural-urban dichotomy, however, masks great disparities within both cities and rural areas. The incidence of disease can vary widely, both within cities and outside them, reflecting differences in socioeconomic status, living conditions, proper nutrition, and access to healthcare and other services (Timaeus and Lush 1995). The incidence of asthma, for example, is high in New York City, but it is particularly high in Harlem—it afflicts 30 percent of children under age twelve—where children have less access to intervention programs and suffer from higher exposure to secondhand smoke than children in other neighborhoods (Nicholas et al. 2005). The environmental justice literature makes a clear case that health risks are unevenly distributed, a controversial issue to which we return later in this chapter. What most experts agree on is that both cities and rural areas can be made healthier than they are now, or at the very least that conditions to improve overall health should be promoted.

HEALTHY CITIES MOVEMENT

While the idea of making cities healthier places to live is not new, over the past twenty years hundreds of municipalities have created "healthy city" initiatives. The basic principle of healthy cities is that health depends on more than good medical care; a healthy environment and nurturing community are key elements as well. The term "healthy cities" originated in 1984 with the Health Toronto 2000 Convention, sponsored by the World Health Organization. The healthy city was defined as "one that is continually developing those public policies and creating those physical and social environments which enable its people to mutually support each other in carrying out all functions of life and achieving their full potential" (Kenzer 2000). At a subsequent conference, the Ottawa Charter for Health Promotion (1986) outlined the same basic principles, emphasizing the need to improve environmental conditions and social well-being in order to enhance overall health. The charter outlines the basic prerequisites for health as "peace, shelter, education, food, income, a stable ecosystem, sustainable resources, social justice and equity" (Duhl and Sanchez 1999, 40).

The use of the terms "sustainable" and "ecosystem" point to the convergent ideas of development and environmental issues in the 1980s as articulated in the philosophy of sustainable development. Giddings et al.'s (2002) conception of sustainable development (see Chapter 6) ably makes this connection between human and ecosystem health. Indeed, a stalwart advocate of healthy cities argues that healthy communities must be environmentally as well as socially sustainable (Hancock 2000).

The goal of healthy cities, however, does not always translate into action. Municipalities that have healthy city programs tend to focus on more mundane, although important, preventive activities such as reducing smoking and alcohol consumption or improving nutrition and increasing exercise. Copenhagen, for example, promotes classes in smoke cessation, diet, and exercise, the prevention of alcohol abuse, and volunteer groups that encourage citizen participation (Copenhagen Health Administration 2004). Dublin's healthy city plan fits more closely with the principles of healthy cities. Like Copenhagen, it promotes initiatives to reduce smoking and alcohol consumption, but it also advocates "environmental" activities, most of which involve volunteering for city cleanup (Dublin Healthy Cities 2004). Although the principles of healthy cities are stated in the plans, it is unclear how these healthy city activities differ from basic public health campaigns.

The idea of healthy cities, if not the practice, takes a systems approach to health, recognizing the links between human and ecological well-being. As with the concept of sustainable development, this is both a strength and a weakness. The holistic approach to health makes sense, given that a person's health is related to a number of factors. Someone in good physical condition can suffer ill health from poor air quality, or even from loneliness. Lack of green space may discourage walking and exercise. People who lack access to clean water, food, housing, and sanitation are certainly more likely to suffer ill health than those who don't. The healthy cities initiative promotes such systematic thinking. On the other hand, because of the breadth of the approach, it is difficult to get a handle on what the healthy city is and how it can be realized. Like sustainable development, healthy cities can mean different things to different people. Hancock et al. (1999) have noted this shortcoming, arguing that the concept of the healthy city is well understood but that evaluating the practice of the healthy city has not been well defined. They argue that a core set of indicators, developed in consultation with the community and users of the information, is a critical step.

The idea has nevertheless caught on. The European office of the World Health Organization reports more than a thousand healthy city initiatives in wealthy countries and nearly the same number in poor countries (Werna et al. 1999; World Health Organization 2004a). In the latest phase of the program, the emphasis for European cities is on integrating urban planning with health planning. Since both types of planning are usually the domain of local governments, such an approach is seen as an effective means to improve the overall health and well-being of the community. Reflecting the demographics of Europe, the emphasis of phase 4 of the plan is on healthy aging. The third goal, perhaps in response to early criticism of the healthy cities movement, is to develop methods of assessing the impact on health. This is an interesting turn, since the healthy cities movement has tended to emphasize planning, prevention, and holistic methods of improving conditions for health rather than measuring impacts on health per se.

In Latin America, the healthy *municipios* movement has stressed the need for access to basic urban services, such as water, electricity, and sanitary sewers, and the need for poverty reduction and better urban governance (Montiel and Barten 1999). Other healthy cities projects in poor countries have emphasized the need for cooperation among local governments, international agencies, and other stakeholders to ensure that a system of health improvement is a priority (Burton 1999; Werna et al. 1999). Healthy cities programs, because of their inclusive approach to improving health, also have the potential to bring together healthcare workers, environmentalists, social workers, and government officials. A study from Leon, Nicaragua, demonstrates how the healthy municipio program did just that, and how it created a critical framework for dealing with the devastation wrought by Hurricane Mitch in 1998 (Montiel and Barten 1999). The healthy city idea is relatively new, even though it borrows many concepts from the public health campaigns of the late nineteenth century (Condran and Crimmins-Gardner 1978; Melosi 2000). It may take some time before the practices match the lofty ideals. Like the philosophy of sustainable development, the healthy cities principle is laudable for its attempt to synthesize the health of environment and society instead of trading one for the other. An integrated approach to environmental and human health can improve overall health in cities, and probably at lower costs than existing healthcare methods, since it emphasizes prevention more than cure (Smith et al. 1998).

AUTOMOBILE CITIES ARE BAD FOR YOUR HEALTH

No technology has had such a profound effect on cities as the automobile. Wherever the car has been adopted on a large scale, the shape and function of cities is altered. Because of the mobility they provide and the space they require, cars tend to lead to low-density development, or sprawl. Dependence on cars for transit results in huge energy consumption, compared to using public transit, walking, or cycling (see Chapter 4). Driving cars also has important health consequences from pollution of air, accidents, and decline in exercise. Automobiles, directly or indirectly, are a major health hazard in cities (Frumkin et al. 2004).

Asthma is the most chronic disease afflicting children today, affecting fifty-four of every thousand children. The disease is triggered by genetic and environmental factors, including particulate matter and smog. The National Center for Health Statistics recorded 10.4 million outpatient visits and 1.2 million emergency room visits in 2000 (National Center for Health Statistics 2004). A 1997 survey showed that 2.2 million Canadians—12 percent of children and 5 percent of adults—have asthma. It is alarming that in Canada and other wealthy countries asthma rates have increased for children over the past twenty years, for reasons that are not entirely clear (National Asthma Control Task Force 2000). The World Health Organization shows wide deviation in asthma rates for children, from zero percent in Papua New Guinea to 30 percent in Australia. Affluent populations tend to have higher rates of asthma than poorer populations (World Health Organization 1995). Pinpointing the cause or causes of asthma is very difficult, but epidemiologists point to the importance of environmental factors. During the 1996 Olympic Games, the increased use of public transit resulted in a 22 percent decline in traffic counts, a 28 percent decline in ozone, and a 41 percent decrease is acute asthma cases requiring care (Cummins and Jackson 2001). Studies of asthma rates after the reunification of Germany showed that asthma rates were higher in Munich, which had more automobile traffic, while bronchitis rates were higher in Leipzig, which had more heavy industrial pollutants (World Health Organization 1995). A study of environmental conditions in France, Switzerland, and Austria shows that more people die prematurely from diseases related to air pollution (21,000) than car accidents (10,000) every year (United Nations Environment Programme 2002).

While the epidemiology of asthma and other diseases remains murky, no one wants to breathe polluted air. Visible pollutants tend to attract most concern in cities, and smog, or ground-level ozone, tops the list. The hazy orange pollutant is the result of photochemical reactions. Ultraviolet radiation from the sun reacts with nitrous oxides, hydrocarbons, and other emissions to produce the smog that blankets so many of the world's cities. Cars are not the only source of smog, but they are a major contributor (emitting up to half of the volatile organic compounds and nitrous oxides and 90 percent of the carbon monoxide found in smog), causing regulators, especially in smog-prone California, to impose increasingly strict emission controls for automobiles (U.S. EPA 1990). Nevertheless, many parts of the United States, including California, do not meet national ambient air quality standards. In 2002 146 million people, or slightly more than half the country's population, lived in counties that did not meet national standards (U.S. EPA 2004a).

In 1970 the U.S. Congress passed the Clean Air Act, an important precedent for combating air pollution, but it set goals that, to this date, have not been attained. The Clean Air Act authorized the U.S. Environmental Protection Agency to establish national rather than regional air quality standards. Although the EPA set the national standards, states were required to draft state implementation plans to achieve these goals. The EPA set minimum standards, but states are permitted to set stricter goals. Initially the act settled on 1975 as the target date for compliance, but, because not all states had met the standards by then, the act was amended in 1977 to set new dates for compliance. In 1990 the act was amended again to address problems of acid precipitation, ground-level ozone (smog), stratospheric ozone depletion, and air toxins, as well as to extend the deadline for compliance (the air quality standards have still not been met). The six major air pollutants measured are nitrous oxides, ground-level ozone, sulfur dioxide, particulate matter (PM10 and PM2.5), carbon monoxide, and lead.

The most significant amendment to the Clean Air Act in 1990 allowed for market-based approaches to reducing emissions, sometimes called emission trading. This gives businesses and utilities the right to trade, buy, or sell pollution allowances rather than having to meet blanket EPA requirements. The business community generally favors this system, as it provides more flexibility and uses market rather than regulatory forces to reduce pollution. Sulfur dioxide emissions, the precursor of acid precipitation and an irritating pollutant that causes respiratory illnesses, have been reduced by more than a third over the past thirty years, suggesting that the system works (U.S. EPA 2000). The trading plan does set limits, based on a formula of sulfur dioxide emissions and the output of power plants (which produce more than 65 percent of SO_2) between 1985 and 1987. In the second phase of the program, the amount of SO_2 allowances was reduced by half. By 2010 total sulfur dioxide emissions from major utilities will be capped at 8.9 million tons, down from 17 million tons in 1980 and 15.6 million tons in 1990. New utilities are not granted credits, so they must purchase them at annual auctions.

The trading system has advantages. It uses a market-based system that businesses can understand. Also, it puts a dollar figure on what was previously considered an "externality" (free air to absorb pollution). But critics argue that the system allows businesses to buy their way out of a problem, and that fragile ecosystems are treated the same way as resilient ones because all traders are treated equally, regardless of their location. The EPA recently began a trading system for mercury, a highly toxic chemical that is more lethal in some places than others, depending on the location of the emitter in relation to population densities and type of ecosystem. While sulfur dioxide is harmful, mercury can be lethal, prompting many to argue that mercury emissions should be regulated by the government rather than defined by the market (Natural Resources Defense Council 2004).

The good news is that air quality is improving in the United States. In the past twenty years emissions decreased for all major pollutants. Lead emissions declined by an impressive 93 percent, mainly the result of phasing out lead in gasoline beginning in 1973 (surprisingly, many European countries did not ban leaded gasoline until after 2000). Air quality also showed an improvement from 1983 to 2003 for all of the six pollutants, but eight-hour concentrations of ozone increased by 4 percent over the past ten years (U.S. EPA 2004a). The latest data show a decline in smog for 2003, although this may be related to favorable weather conditions (U.S. EPA 2004b). Ozone levels remains high, especially in California, the Northeast, and around major metropolitan areas. Despite advances in emission controls for vehicles, Americans drive more miles than they did twenty years ago. Increased congestion and stop-and-go traffic also make emissions worse, since automobile emission controls are most effective at cruising speeds.

Southern California, despite strong regulation and progressive air quality practices, continues to have poor air quality that does not meet the national ambient air quality standards. A growing population coupled with perfect physical conditions for smog—plenty of cars, sunlight, and temperature inversions—makes improving air quality a major challenge. Other cities suffer from similar circumstances. Like Los Angeles, Mexico City, with perhaps the worst air quality of any city in the world, is surrounded by mountains, which trap warn air underneath cold, reducing vertical mixing, or the ability of air to clean itself through dilution. Exhaust from Mexico City's 4 million vehicles, most of which are old and in poor condition, is the major contributor to smog. Ozone levels are above World Health Organization standards more than three hundred days a year (U.S. Department of Energy 2004).

In 1989 Mexico City began a "no driving day" program that prohibited the use of cars on the basis of license plate numbers. The program immediately reduced the number of cars on the roads by 20 percent, increasing speeds by eight kilometers per hour, reducing average trip time by five minutes, and increasing subway use by nearly 7 percent. Impressed by this success, the local government decided to make the program permanent. Ironically, however, the program made congestion and air pollution worse. Once Mexico City residents understood that the program would be permanent, it encouraged them to buy more cars so that they would have a vehicle to drive everyday. For those who could not afford to purchase a second car, many simply shifted their automobile-related tasks from a banned day to another day, with the result that the same number of miles were driven but were concentrated during permitted driving days. The program also encouraged people to buy inexpensive second cars. Normally an exporter of old vehicles to the countryside, the capital city began importing them after the program began, putting more older, polluting cars on the roads (Eskeland and Feyzioglu 1997). Mexico City's experience shows that well-intentioned plans to improve environmental conditions can have unforeseen consequences. At the same time it demonstrates that municipalities are willing to take desperate measures to reduce automobile use.

CARS, WALKING, AND CYCLING

Obesity has reached epidemic proportions in the United States. Self-reported height and weight measurements indicate that obesity rates have risen from 12 percent in 1991 to 20 percent in 2000. Clinical measurements indicate that 31 percent of American adults were obese and 64 percent were overweight in 2000 (Pucher and Dijkstra 2003). Poor diet and lack of exercise are major contributors, but recent studies have linked urban design to obesity. A recent report published in the *American Journal of Health Promotion* showed that people living in sprawling suburbs are more likely to be overweight than people living in dense urban settings. Geauga County, Ohio, outside Cleveland, has the highest sprawl index in the country, and on average its citizens were six pounds heavier than residents of Manhattan. The study found that people living in sprawling counties not only weighed more but had higher rates of hypertension and walked less than residents living in compact cities (Ewing et al. 2003a). Rates of Type-II diabetes are also on the rise, and that disease, normally associated with adulthood, is now occurring in children. Increasing rates of overweight and obese children are likely to create a new childhood diabetes epidemic (Botero and Wolfsdorf 2005; Perdue et al. 2003).

Numerous factors may help to explain the difference in the number of overweight and obese persons in the sprawling suburbs and the number in densely built-up downtowns, but "walkability" is a major one. In a typical sprawling suburb, most residents drive to perform errands, while in densely settled urban cores, people are less likely to travel by car for all trips. The 2001 National Household Travel Survey shows that Americans use cars for 66 percent of all trips less than a mile and 89 percent for trips between one and two miles. Personal vehicles accounted for 87 percent of all trips, while walking accounted for 8.6 percent and public transit for a meager 1.5 percent (Table 5.2). Remarkably, 91 percent of commuting trips, where public transit is supposed to work best, are done by private vehicle. The average driver spends fifty-five minutes behind the wheel every day and travels twenty-nine miles (Bureau of Transportation Statistics 2001).

Transportation statistics in other countries are quite different. In Canada, 10 percent of urban trips are done on foot. In France and Italy,

TABLE 5.2. Percentage of Trips by Transportation Type, U.S., 2001

Mode	Percentage
Personal vehicle	86.6
Vehicle with single-person occupant	37.6
Vehicle with multiple-person occupants	48.9
Public transit	1.5
School bus	1.7
Pedestrian	8.6
Other	1.7

Source: Bureau of Transportation Statistics (2001).

24 percent of all trips within cities are done on foot, and for Sweden the figure is 29 percent. The same differences show up in cycling rates. Only 1 percent of all urban trips in the United States are done by bicycle, compared to 4 percent in France, 12 percent in Germany, and a whopping 28 percent in the Netherlands, where excellent bicycle trails, tolerance of cyclists, and flat terrain all encourage bicycle use (Pucher and Dijkstra 2003). The high cost of car ownership and fuel in Europe also discourages the use of cars for short distances (Table 5.3).

Pedestrian accidents happen more often in the United States than in European countries. An American pedestrian is three times more likely than a German and six times more likely than a Dutch pedestrian to be killed by a car. This is the case on both a per-trip and a per-kilometer basis. When Americans bicycle, they are twice as likely to be killed as a German cyclist and three times as likely to be killed as a Dutch cyclist,

TABLE 5.3. Number of Registered Private Passenger Vehicles Per Capita for Selected Countries

Country	Registered Cars (thousands)	Cars per 1,000 Persons
Sweden	3,630	408
Belgium	4,239	424
Netherlands	5,633	368
Canada	13,182	455
Great Britain	24,306	411
France	24,900	429
West Germany	40,499	494
United States	148,500	571

Source: Transportation Research Board (2001).

also on a per-kilometer and per-trip basis. In the United States, for every kilometer traveled, pedestrians are twenty-three times, and cyclists twelve times, more likely to be killed than people traveling by car (Pucher and Dijkstra 2003). Another study found that low-density, sprawling counties in the United States have significantly higher fatality rates for pedestrians and car occupants than do compact, densely settled metropolitan areas. In 2001 Manhattan had 4.42 traffic fatalities per hundred thousand people, while Geauga County, Ohio, had a fatality rate of 20.90 (Ewing et al. 2003b).

The experience of Europe shows that accident and death rates for pedestrians and cyclists can be reduced through careful planning, education, and enforcement of traffic laws. In Germany pedestrian fatalities dropped by 82 percent, and cyclist fatalities by 64 percent, from 1975 to 2001. Remarkably, the number of cyclist fatalities declined even though ridership increased by 50 percent. In the United States over the same period, pedestrian fatalities decreased by 27 percent, but the decline was due almost entirely to a drastic reduction in the number of children riding bicycles (Pucher and Dijkstra 2003). Understandably, the decline in child cycling is a response to increased perception of the dangers posed by automobiles, but it is one of many factors that has resulted in increased inactivity and higher obesity rates in children (Cummins and Jackson 2001).

Death rates from car accidents, on the other hand, have decreased over the past twenty years but still remain high. In the United States in 2002, thirty-eight thousand people died in car accidents, down from fifty thousand in 1978, while the number of licensed drivers increased by 80 percent over that period. Nearly 2 million people were injured in 4.3 million car crashes in 2002 (U.S. of Transportation 2002a). The fatality rate dropped from 5.5 per 100 million vehicle miles traveled in 1966 to 1.5 per 100 million in 2002. Mandatory seatbelt laws, as well as better safety equipment in cars, have improved survival rates in car crashes, but the number of deaths is still alarming. Motor vehicle accidents are the leading cause of death for persons age two to thirty-three (U.S. Department of Transportation 2002b).

Wealth generally means more cars, and with cars come deaths from traffic accidents. In 2000 there were 126,000 fatalities from car accidents in OECD countries. This figure is down from 157,000 in 1980, even while population has grown over that period. Only in the newly expanding economies of Korea, the Czech Republic, and Poland did the number of fatalities increase over that period, owing to higher rates of car ownership. The average fatality rate for all OECD countries in 2000 was 12.5 per hundred thousand population, but Britain, Sweden, Norway, and the Netherlands have dropped their fatality rates to below 7 (OECD 2002). For the United States the figure was 14.9 and for Canada 9.6. Korea had the highest fatality rate in 2000, with 21.8 per hundred thousand. This figure, however, does not take into account the number of cars or the number of miles driven. For these reasons, the OECD also publishes the number of fatalities per billion vehicle kilometers traveled. Canada and the United States both have low figures of 9.45, while Sweden boasts the lowest figure—8.34. Turkey had the highest rate, 119.82, in 2000, although the number of fatalities per hundred thousand was relatively low at 7.6 (OECD 2002).

Although city dwellers pay higher car insurance premiums, the death rate from car accidents is higher in rural than in urban areas. Data from Minnesota, for example, show that only 27 percent of traffic deaths occur in cities. Nationally, 40 percent of fatal crashes occur in urban areas. Gender and age are significant factors, as parents of teenage boys know when they look at their car insurance bills. In Minnesota 67 percent of those killed in traffic accidents are male. Although fifteen- to twenty-four-year-olds make up only 17 percent of licensed drivers in the state, they are involved in 30 percent of all vehicle accidents. Forty percent of deaths from car accidents are in the fifteen-to-thirty-four age cohort (Minnesota Safety Council 2002). National data show that male drivers age sixteen to twenty have the highest traffic fatality rate of any age and sex cohort, and more than twice the death rate of females in the same age group (U.S. Department of Transportation 2002a).

Many of these deaths are preventable. National data for the United States show that in 2001 more than half of those killed in automobile accidents were not wearing safety belts. Alcohol was related to 41 percent of fatalities, a slight decrease from 47 percent in 1992 (U.S. Department of Transportation 2002a). Improved driver education, increased attention to drunk driving (notably by Mothers against Drunk Driving), and stricter blood-alcohol levels (0.08 g/dl) have helped to reduce these figures. Another way to reduce your chance of dying in a car crash is to drive during daylight hours. Fatality rates continue to be highest from midnight to 3:00 A.M., when 76 percent of all fatal crashes involve alcohol (U.S. Department of Transportation 2002a).

Some cities are more dangerous than others. Table 5.4 shows traffic fatality rates for American cities in 2002. New York has the most deaths in absolute terms, but its traffic fatality rate (per hundred thousand

TABLE 5.4. Traffic Fatalities and Fatality Rates for Selected U.S. Cities, 2002

City	State	Fatalities	City Population	Fatality Rate (per 100,000 population)
Louisville	KY	69	251,399	27.45
Knoxville	TN	44	173,661	25.34
Tampa	FL	71	315,140	22.53
Orlando	FL	38	193,722	19.62
Miami	FL	69	374,791	18.41
Kansas City	MO	78	443,471	17.59
Atlanta	GA	73	424,868	17.18
Phoenix	AZ	232	1,371,960	16.91
Minneapolis	MN	14	375,635	3.73
Irvine	CA	6	162,122	3.70
Spokane	WA	7	196,305	3.57
Fort Wayne	IN	6	210,070	2.86
Overland Park	KS	4	158,430	2.52
Glendale	CA	5	199,430	2.51
Arlington	VA	0	189,927	0.00
National City Average		40.47	—	9.43

Source: U.S. Department of Transportation (2002a).

population) is one of the lowest in the country. Yet it does have the highest number of pedestrians killed in traffic crashes and one of the highest pedestrian fatality rates in the country. San Francisco had less than a tenth of the traffic fatalities of New York, but more than half the deaths were of pedestrians. Given that these are walking cities, the figures for pedestrian deaths may come as no surprise, since more walkers and fewer drivers mean more chances of a pedestrian fatality. Overall, the traffic fatality rates for New York (4.4) and San Francisco (5.5) are about half the average for all U.S. cities (9.4). Louisville, Kentucky, had the highest traffic fatality rate per hundred thousand population in 2002, followed by Knoxville, Tennessee, and Tampa, Orlando, and Miami, Florida.

One way to reduce the chance of death or injury while traveling in the United States is to take public transit. In 2000, 295 people died and fifty-seven thousand were injured while on public transit. Fatality and injury rates for public transit are expressed as the number per 100 million passenger miles. The fatality rate for all forms of public transit in 2000 was 0.697, while the fatality rate for automobiles was 1.5

per 100 million vehicle miles. Data show that buses, at 0.506 fatalities per 100 million passenger miles, were safer than rail (0.822). The number of public transit fatalities from 1992 to 2000 has remained fairly constant, averaging 285, with a peak of 320 in 1994. The number of passenger miles, by contrast, has been steadily increasing, from 34.1 billion miles in 1992 to 42.3 billion in 2000 (Federal Transit Administration 2000). But public transit ridership still pales in comparison to the 1.6 *trillion* vehicle miles logged by passenger cars, amounting to less than two-tenths of 1 percent of all miles traveled (Bureau of Transportation Statistics 2001). Shifting even a fraction of those trips from the automobile to public transit, walking, or cycling, would have significant positive effects on the health of the American population.

TRANSPORTATION PLANNING AND THE THREE D'S: DENSITY, DIVERSITY, AND DESIGN

Transportation planners note that the three D's—density, diversity, and design—are critical factors in determining transportation mode (Cervero and Kockelman 1997). Higher densities, mixed-land uses, and pedestrian- and cycling-friendly designs encourage people to walk or cycle for short trips rather than drive (Greenwald 2003). The design philosophy of new urbanism (see Chapter 6) attempts to make neighborhoods more friendly for walking and cycling by creating roads and paths that encourage the three D's, and evidence suggests that it works (Lund 2003). While other factors, such as topography, weather, and demographic characteristics, are also important, these are difficult if not impossible to control or modify (Cervero and Duncan 2003). Careful planning, on the other hand, can create a built environment that is more amenable to walking, cycling, and public transit. Critics suggest that American cities have largely failed to promote the three D's.

Compact, high-density cities tend to encourage more walking, cycling, and public transit use than do sprawling cities and suburbs. For one thing, origins and destinations tend to be closer, and people are more likely to walk or cycle short distances. High densities increase the efficiency of public transit, usually measured in passengers per car mile (or passengers per unit kilometer), since more people, the transit market, are found in less space and are likely to travel shorter distances than in low-density areas. Research in the United States shows that residential density must be above three dwellings per hectare in order to maintain a 5 percent use of public transit (Transportation Research

Board 2001, 37). Compact cities also tend to have less parking, or higher-priced parking, than low-density areas, which discourages car use (Cervero and Kockelman 1997). But large subsidies for parking in central cities, usually paid by employers, can undermine any impact that densities might have on parking costs.

Population densities of American cities are generally lower than their European counterparts, especially in the central cores. The median density for central cities in the United States is twenty-four hundred per square kilometer compared to thirty-four hundred for western Europe (Transportation Research Board 2001). Although western European cities have also suburbanized, they have not done so to the extent of North American cities. The multinucleated settlement and the development of edge cities are very different in western European cities that retain a traditional economic and population core in the center (see Chapter 1 on morphology). The European model is better suited to transit systems, at least as they are traditionally designed, radiating out from the core to the periphery, rather than trying to serve multiple cores. Very low population densities in North American suburbs also make public transit inefficient.

For western Europe, the timing of growth helps explain the high densities and, in turn, the higher transit use and noncar travel compared to the United States. The cores of many European cities were built long before the arrival of the automobile and so were built with people, walking, and public transit in mind. As a result, the density of the built environment is higher, streets are narrower, and buildings are closer to the street, to serve walkers rather than parking. But a host of other factors help to explain the higher transit use in Europe compared to North America. Population growth, for example, has been far greater in North America than in western Europe. Since 1950 the U.S. population has more than doubled, adding 130 million people, while the population of France, Germany, and Great Britain combined has grown by only 40 million, or 25 percent. In other words, the United States has had to absorb more population growth over the past fifty years than has western Europe, and at a time when the automobile was increasing as a popular means of travel on both sides of the Atlantic. In addition, the population of the United States is younger than that of western Europe, and people with young families are more likely to live in suburbs than older people without children at home (Knox and Pinch 2000). Affluence in North America during the post–World War II boom contrasted markedly with economic conditions in war-torn Europe, where home or car ownership was beyond the means of many. Differing forms of government, road financing and construction, fuel taxes, and transit planning have all contributed to different kinds of cities and transportation methods. The fact that central cities continue to house the elite in western Europe and in some Canadian cities rather than the poor and immigrants, as in many American cities, also helps to explain support for public transit and differences in transit use (Transportation Research Board 2001).

American travelers often remark on the "walkability" of European cities (Rybczynski 1995). Besides higher densities, the diversity of land uses in older western European cities tends to encourage walking and cycling. The same is true of certain American cities, such as Portland, Santa Monica, New York, and San Francisco (Portney 2003). By mixing residential land uses with shopping, workplaces, parks, recreational spaces, places of worship, and so on, distances between origins and destinations tend to be shorter. If land uses are separate from one another, encouraged by zoning laws that have attempted to separate nuisance land uses from residential areas, distances tend to be greater, encouraging people to drive rather than walk. Other elements of design and diversity tend to make some places more attractive for walking than others. Buildings set close to wide sidewalks, shaded streets, benches for sitting, and diversity of services tend to promote walking. Quite simply, having something interesting to see while walking encourages walking. The so-called "blandscapes" of most suburban subdivisions do not tend to interest walkers (Relph 1987). Most people would rather hurry by familiar suburban scenes to get to their destinations, taking little pleasure in the travel itself.

Suburban morphologies are a product of the automobile age and reinforce dependence on cars. The curvilinear design and cul-de-sac are meant, in part, to reduce through traffic, but ironically they encourage more car use. Circuitous routes mean that even traveling to a friend's house discourages walking or cycling. In many suburbs developers have stopped putting in sidewalks, knowing that few people walk or cycle. This further discourages walking and cycling, or makes it more dangerous because pedestrians and cyclists have to share the road with cars. The streets themselves encourage high speeds because of their width, originally designed to accommodate fire and safety vehicles. Since there is plenty of off-street parking in driveways and garages, fewer cars on the road also encourage higher speeds (Kunstler 1993).

Public transit systems evolved to transport people to the central core of cities, where most businesses and workplaces were typically located.

The changing morphology of North American cities has altered the classic urban core and suburban residential area layout. In 1990 45 percent of all jobs were located in central cities, a drop from 70 percent in 1950 (Mieszkowski and Mills 1993). Transit ridership in the United States dropped from 115 rides per capita in 1950 to thirty-five in 1990 (Transportation Research Board 2001). As fewer people traveled to the downtown core, public transit systems became less useful to commuters. The extensive electric streetcar networks in American cities, which peaked at seventy thousand kilometers in 1920, began to decline as more flexible modes of transit, including buses and cars, replaced the fixed-rail transit lines. Much has been made of the "conspiracy" of General Motors, Standard Oil, and National City Lines in buying the trolley systems, tearing up the tracks, and converting to their own bus systems. While there certainly was collusion (which resulted in a meager $5,000 fine for General Motors in 1949), the companies merely accelerated a process already under way (Adler 1991; Jackson 1985). Restricted fare increases, aging rolling stock, an image of being old-fashioned, and the fixed nature of streetcars all contributed to decreased ridership in the 1920s (although ridership enjoyed a brief peak during World War II due to rubber shortages for civilian use) (Transportation Research Board 2001). Some cities, Toronto among them, kept their streetcars not because of clever foresight but because they had purchased the private streetcar companies and made them public entities, an investment they did not wish to tear up (Armstrong and Nelles 1986). Streetcars and buses have now traded places in the United States. Buses are perceived as transit for the poor, while streetcars and other light rail systems are seen as modern, more agreeable forms of public transportation. Cities have responded by building new streetcar lines, even though buses are cheaper and fares pay more of their operating costs (Wachs 1996).

Data from all corners of the globe show that cars are more popular than ever as a means of transit. In 1998 700 million vehicles were in use worldwide, an increase of 145 million since 1990. The number of passenger vehicles in use reached 517 million in 1998, up more than a 100 million since 1990. The largest increases in passenger vehicles over that period occurred in Asia, which added nearly 40 million vehicles over the eight-year period (OICA 1999, 95–96). A growing Chinese economy is likely to fuel future growth. From 1990 to 1999 the number of passenger vehicles in use grew from less than a million to more than 5 million (U.S. Department of Commerce 2001).

Urban planning for more private vehicles has responded by building low-density, sprawling suburbs and shopping malls surrounded by plenty of parking, and by constructing highways, even if housing has to be destroyed or green spaces intersected in the process. An amazing amount of time, energy, and money has been spent to accommodate the car and driver. Yet not everyone wishes to drive all the time, as the success of pedestrian-oriented development illustrates (Rybczynski 1995). Old Town Pasadena, just northeast of Los Angeles, is crowded with pedestrians on any given night, even if people have to drive to get there and pay parking fees. Although some may call it nothing more than an outdoor shopping mall, it provides a pedestrian experience that people are willing to pay for. Even traditional enclosed shopping malls, some of which are designed to look like city streets, appeal to the desire of people to walk and see other people in a convivial setting (consider the contrast in behavior, on both sides, when someone cuts off another car in traffic, as opposed to steeping in front of them on a sidewalk) (Gladwell 2004). Meeting the latent demand for walking, the original human means of locomotion, should be part of any urban design, given the benefits to physical and mental health.

ENVIRONMENTAL JUSTICE AND HEALTH

Not all parts of the Los Angeles metropolitan area are amenable to walking, which explains the large crowds that gather in Old Town Pasadena, Santa Monica, Venice Beach, and other pedestrian-friendly locations. Cities are heterogeneous in their characteristics; they display a certain patchwork of land uses and functions that make some areas more pleasant and inviting than others. The clustering of land uses, such as industry, shopping, and housing, is the result of a multitude of factors, including market forces, regulation, technology, and urban planning. The result is that some parts of the city enjoy more amenities and some have more disamenities. The environmental justice movement, which has gained momentum over the past twenty years, tries to redress the disproportionate number of disamenities often found in disadvantaged communities. The next frontier of environmental justice scholarship will examine the distribution of amenities as well, such as the ability to walk in pleasant surroundings.

The focus on disamenities is understandable because of the real or perceived risk of living close to a disamenity, such as a factory spewing toxics into the air or an abandoned industrial yard with hazardous waste on site. Aesthetics may play a part, but the vast majority of environmental justice movements have focused on the *health* risks of living close to unwanted land uses (Bullard and Johnson 2000). The modern environmental justice movement owes its origins in part to the Environmental Planning and Community Right-to-Know Act (EPCRA) passed by the U.S. Congress in 1986. This act, which made public the location and activities of facilities that release toxics, was a response to the disaster in Bhopal, India, in 1984. In one of the worst industrial accidents on record, a plant owned by Union Carbide exploded in December 1984, killing at least two thousand people and injuring many thousands more. What alarmed Indians and the international community is that nearby residents had no knowledge of what the plant was producing or the possible dangers of living close by. When another accident happened at a sister plant in West Virginia, Congress passed the EPCRA.

Most significantly, EPCRA gives the public the right to know what toxics are being released by facilities in their communities. The legislation requires that most facilities report levels of toxic emissions annually to the EPA, which was charged with creating a publicly accessible database called the Toxics Release Inventory (TRI). This was the first database of its kind in the world, and it has been used extensively in environmental justice research. Some have called the EPCRA the most important piece of environmental legislation ever passed because it puts information, and thus power, in the hands of the public (Fung and O'Rourke 2000). TRI information has been an effective means of "shaming" polluting facilities into reducing toxic emissions, relocating, or shutting down altogether.

Early environmental justice studies showed that race rather than income was the best explanatory variable for the location of hazardous facilities (United Church of Christ and Commission for Racial Justice 1987). This has led some scholars and activists to declare that hazardous facilities are deliberately located in minority communities, a process described as environmental racism. Others have argued that environmental racism is more than deliberate siting. The mere presence of toxic facilities in predominantly minority communities, the argument goes, is proof enough that environmental racism is at work. Although most probably the result of institutional and other forms of racism, the inequitable or unjust result, such advocates suggest, is sufficient to declare it environmental racism and place the burden of proof on polluting industries to demonstrate otherwise (Bullard 1994a; Bullard 1994b; Bullard and Johnson 2000; Pulido 2000; Pulido et al. 1996). Environmental racism, intentional or otherwise, has further identified environmental justice movements with issues of minority rights. For this reason campaigns for environmental justice are most often associated with minority communities and grassroots activism.

Certainly the TRI database and other databases provided by the EPA have been important ammunition in the environmental justice movement. Bullard and Johnson (2000), however, argue that the environmental justice movement began with grassroots activism, not with government, academia, or middle-class environmental organizations. High-profile legal challenges have allowed community groups to bring national attention to local issues of environmental justice. Grassroots organizations have also been able to participate on national boards and agencies to achieve results. Operating at both local and national levels has been an effective strategy.

The ability of the environmental justice movement to achieve social justice stems in large part from a 1994 presidential executive order requiring that all federal agencies develop environmental justice strategies as part of their mission. This ensured that environmental justice would remain in the limelight, and community activists have seized on this commitment. Citizens against Toxic Exposure, for example, used environmental justice to argue for the relocation of nearly four hundred homes adjacent to a superfund site in Florida, forcing the EPA to amend an earlier plan to relocate only a fraction of the homes. The campaign was initiated by Margaret Williams, a retired schoolteacher, who feared for the health of residents living near a wood-treating site that local residents called "Mount Dioxin" (Bullard and Johnson 2000). In Louisiana, a group called Citizens against Nuclear Trash (CANT) effectively argued that a decision by Louisiana Energy Services and the Nuclear Regulatory Commission to locate a uranium enrichment plant in their community amounted to environmental injustice. A three-judge panel agreed and declared that the agencies practiced racial bias in their siting decision (Bullard and Johnson 2000). Such high-profile cases can lead communities to be proactive. In the city of Commerce, a predominantly Latino community located in eastern Los Angeles County, a newly elected city council refused in 1997 to allow the completion of a multimillion-dollar recycling facility, citing it as a case of environmental injustice. The following year the council passed an ordinance prohibiting any waste-processing facility within the city and incorporated explicit environmental justice language into the text (Boone and Modarres 1999).

Not all cases of environmental justice are pursued by environmental or community groups, and not all cases are successful (Bullard and Johnson 2000). High-profile sources of pollution or toxicity are prime targets that activists are prone to exploit. Fung and O'Rourke (2000) argue that the availability of the TRI database has created a form of "populist maxi-min regulation" through which pressure on polluting industries comes from ordinary people who spend the maximum time and effort on exposing the minimum (or worst) performing facilities. By generating blacklists or top-ten lists, community groups have been effective in reducing overall toxic releases. The authors argue, furthermore, that this kind of strategy has been far more effective than the typical "command and control" regulations the EPA has historically employed. Although they concede that it is difficult to pinpoint underlying causes, the overall release of toxins dropped by 45 percent between 1988 and 1995, after the public release of the TRI database.

Literature on grassroots activism shows that environmental injustice becomes a rallying cry primarily when it is seen to affect *human* health, livelihood, and livability of an area rather than the nonhuman environment. In that sense, environmental justice at the grassroots is more about social justice than it is about concern for the natural environment in its own right. National groups such as the Sierra Club have traditionally tended to focus on how toxins and pollution affect wildlife, ecosystems, and wilderness areas. Over the past decade, the wide umbrella of environmental justice has brought together, to a certain degree, the mostly white, middle-class membership of national environmental clubs with the predominantly minority, working-class members of grassroots urban environmental groups. It has also brought together community groups that usually do not interact. In 1993, for example, Latinos and Hasidic Jews, traditionally suspicious of one another in their Brooklyn communities, marched together across the Williamsburg Bridge to Manhattan to protest the siting of a waste incinerator. The Community Alliance for the Environment, made up of Latino and Jewish community members, managed to delay the construction to the point where it was no longer economical and was therefore abandoned (Gandy 2002).

Obvious point sources of pollution, bellowing smokestacks, for instance, are more likely to be the target of activists than nonpoint sources such as agricultural runoff. Liability is easier to assign to a point source than a nonpoint source of pollution. Given the "populist maxi-min regulation" approach of grassroots organizations, it is simpler to mobilize a protest against a single facility releasing toxic waste than multiple sources that may, in aggregate, be more harmful to human and environmental health.

A second characteristic of the environmental justice movement is a strong focus on disamenities. From a public relations point of view, it is often more practical to highlight disamenities as a form of environmental injustice than lack of amenities. But environmental and health benefits that accrue from vegetation cover, for example, can be interpreted in environmental justice terms (Grove 1996). Indeed, overall human and environmental health might be better served by improving environmental amenities than by focusing on disamenities. Yet the struggle to improve neighborhood environmental services, such as parks or tree cover, can be long and frustrating, while a court decision on a polluting facility can lead to rapid, tangible change, a reality that grassroots activists understandably exploit.

Making concrete links between TRI sites and human health is a complicated exercise, and one that, if it is the sole criterion for environmental justice, could undermine the legitimacy of the movement. Epidemiological studies have great difficulty showing statistical significance between the incidence of disease (e.g., cancer) and environmental conditions (e.g., toxic releases), even though many diseases, like cancer, are known to have links to environmental conditions. Most environmental justice studies have assumed, reasonably, that risk is a factor of distance from a pollution source—the closer someone lives to a toxic site, the greater the health risk. While the model is intuitive, in truth it is difficult to demonstrate. The premise that residential proximity to toxic sites equals risk does not take into account that residents move around the city and that risks might be higher in the workplace than at home. Other factors, such as wind direction, hydrographic conditions, soils, topography, and climate can complicate the neat circular buffers drawn around most toxic sites to delineate risk zones. Exposure times, chronic versus acute releases, and synergistic effects from multiple sources of pollutants can all vary the health risks associated with point sources of pollution (Bowen 1999). Modeling risk is a tricky, complicated process, but living close to a polluting factory must nevertheless be considered more risky to health than not living near one. While it may not be possible to differentiate health risks at the neighborhood level, it should be possible to delineate higher-risk from lower-risk areas.

Environmental justice activists have responded to such criticism by arguing for precaution. Much like the public health campaigns of the nineteenth century, when the science of disease was not entirely understood, precaution can pay off by preventing conditions that lead to disease, even if the process is not entirely clear (Giddings et al. 2002). Current models may not be sophisticated enough to show exact levels of toxics over a prescribed area, but should we wait for such models, or work on simpler premises (such as circular buffers) to reduce exposure near toxic facilities? In many ways, this type of thinking is what the healthy cities movement is about. Rather than focusing on results, healthy cities initiatives promote the process of making a city healthier, arguing that the commitment to continually improving conditions is more important than measuring success. In other words, the commitment to a healthy city is success in itself.

Environmental justice has its roots in the United States, but the ideas and principles are being adopted elsewhere. In October 2001 all European Union countries ratified the Aarhus Convention. The convention stipulates that "every person has the right to live in an environment adequate to his or her health and well-being," and that to achieve this right "citizens must have access to information, be entitled to participate in decision-making and have access to justice in environmental matters" (United Nations Economic Commission for Europe 2003). In language and spirit, the text of the convention mirrors EPCRA as it attempts to provide citizens with information on facilities that may be exposing them to risks. The terms may be different—"environmental equality" and "environmental democracy"—but the intentions are the same. It is too early to tell whether the Aarhus Convention will have the same impact on neighborhood mobilization that the EPCRA has had. Knowledge is power, but only if it is combined with action to highlight inequities. While the lack of amenities is a form of environmental injustice, it is likely that groups will continue to address disamenities because of the perceived or real health risks. The vitality of the environmental justice movement demonstrates that the yearning for a healthy city can be a strong mobilizing force for change.

6   Green Spaces, Green Governance, and Planning

ONLY FROM the window of an airplane does it become apparent that U.S. cities are often heavily forested while surrounding areas are not. On average, more than a quarter of urban land is covered by trees (Nowak et al. 2001). Tree cover is especially apparent on the older fringes of cities, where pressure on land is not as great as in the center, and where time has given trees a chance to grow large canopies. In the spring, the green lawns of the suburbs radiate skyward, while surrounding farm fields, in hues of beige and brown, wait for the farmer's plow. There is already plenty of "nature" in cities even if much of the ecosystem is highly managed. Given the multiple benefits that accrue from green space and urban trees, it is no wonder that many municipalities are making renewed efforts to "green" their cities.

The environmental benefits of green spaces and trees are many, from cooling cities to improving water and air quality (see Chapter 4). Trees also play a role in slowing global climate change. The 3.8 billion city trees in the United States are estimated to store 700 million tons of carbon and sequester 23 million tons per year. The amount stored (not sequestered) is the amount of carbon emitted by the American population every six months. Annual sequestration of carbon by city trees, however, is the equivalent of fewer than six days' worth of emissions nationwide. While city trees sequester only a small portion of the carbon emitted by human activity, they store large amounts of carbon that is better left in the tissue of the tree than released into the atmosphere. Maintaining trees is important for locking up carbon in plant tissue instead of releasing it into the atmospheric carbon cycle. New York City's 5.2 million trees store 1.2 million tons of carbon and sequester thirty-eight thousand tons of carbon every year. The economic value of carbon storage and sequestration in city trees is estimated to be $14.3 billion (Dwyer et al. 2000; Nowak et al. 2001; Nowak and Crane 2002).

Parks, playgrounds, trees, and recreation areas also provide health benefits to mind and body. Although some parks may be seen as disamenities, especially if they are not policed or maintained, most urban dwellers welcome them as amenities and are willing to pay for them. Higher housing values near parks increases property taxes, which help to maintain the parks. Evidence also shows that mature trees, like parks, can add to the value of a home or neighborhood (Crompton 2001).

But green space comes with costs. Mature trees, while beautiful, can also pose a hazard if they topple or their limbs fall. Trees in public places must be maintained with public dollars, a cost that many cash-strapped municipalities are quick to slash when budgets are drawn. Poor selection of tree species can create problems. Some tree species are ill suited to the compacted soils or poor air quality of urban areas. Others create root systems that damage underground conduits, crack sidewalks, or damage building foundations. Maintaining diversity of tree species of different ages is also an important strategy for city arborists. Dutch Elm disease devastated urban forests in the twentieth century partly because of the monoculture practiced (Johnson et al. 2003). The fungus (carried by bark beetles) that infected the trees spread easily from tree to tree. A street lined with old elms can be beautiful, but many cities have recognized the need to adopt an ecosystem approach to urban forestry, planting a diversity of species that are better able to resist pests and disease while providing habitat for birds and other urban critters.

Despite the costs and hazards that come with large trees (or even open space, which can invite crime), most cities are working to increase green spaces because of the environmental, health, and economic benefits they provide. The transition from a complete reliance on "gray infrastructure" to incorporation of "green infrastructure" is part of the renewed commitment. As cities repair aging water and sewerage systems, for example, they are beginning to explore the use of riparian buffers, pervious surfaces, and even artificial wetlands as an ecologically and in some cases economically friendlier means of providing city services than entirely engineered systems (see Chapter 4). The decline of population in central cities has forced municipalities to explore ways to slow the drift of residents to the suburbs, or to lure them back. Urban amenities, including safe, well-maintained parks and open space, are one way to keep people in the city.

PARKS IN THE CITY

Parks have always served as a refuge from the city, even though their design and purpose have changed over time. In the large urban parks of the Victorian era, city dwellers could stroll through an idealized rural landscape and escape the crowds, noise, and pollution of the industrial city. Smaller playgrounds, developed in the late nineteenth and early twentieth centuries, offered places for children to play that were generally safer than the streets, especially as automobiles began to take over thoroughfares. More recently, greenways offer walkers and cyclists a shady respite from busy city streets and also provide habitat and shelter for wildlife. Although parks are part of the urban fabric, they are often perceived as oases of nature separate from the city. One struggle for urban ecology is to convince city dwellers that cities are more than asphalt, concrete, and buildings. Part of the solution is reversing the long-held philosophy that parks are antidotes to the vices of city life.

The earliest green spaces in cities served distinctly economic and practical purposes. The Boston Common was purchased by the city in 1634 so that villagers could graze their livestock there. As a "common," all citizens had the right to use the land for grazing. The system of commons was inherited from Britain, although commons in England were threatened by the enclosures (the fencing in of common land to create large private holdings) beginning in the fourteenth century (Slicher van Bath 1963). Foreshadowing modern parks, the Boston Common was a multipurpose open space. Besides the agricultural function, it was a gathering space for militia and parades, a cemetery, and was used for the gruesome purpose of hanging "unwelcome Quakers" (Platt et al. 1994).

The Boston Common answered the need for grazing fields in a small colonial town. Today it looks like many other urban parks, surrounded by high-density, built-up areas, even if it differed from other city parks in its original purpose. The parks of the nineteenth century were created not for farming or grazing but as a response to the rapidly growing cities in the industrial era. Instead of working rural landscapes, nineteenth-century parks were designed as idealized rural settings, a recreation of countryside, albeit without the smells or sounds of a typical farmstead. Meandering paths invited strollers through parks constructed to look like a utopian countryside of gentle hills, ponds and lakes, footbridges and swans, and the occasional gazebo. These Victorian-era parks were usually built on the edge of cities, where land was less expensive and there was room for the large open spaces that characterized parks in this era (Tuason 1997).

New York's Central Park is one of the more famous examples of the *Rus in Urbe,* or countryside-in-the-city parks. Designed by Frederick Law Olmsted and Calvert Vaux in 1857, the 840-acre park sits at the heart of the financial capital of the world. Olmsted drew on the design and principles of Birkenhead Park (Liverpool), which he visited in 1850, three years after its completion. The first public park built by a municipality, Birkenhead Park displays the classic characteristics of the Victorian parks that followed, with constructed meadows, lakes, and wooded areas crisscrossed with meandering strolling paths. The aesthetics of the park, as well as the democratic ideals of its design, appealed to Olmsted, and these elements were reflected in his design of Central Park and other urban parks in the United States. Fully aware of the cost of park building, Olmsted took great pains to show that parks could pay for themselves by increasing the value of adjacent properties. He learned this lesson from Paxton's Birkenhead Park, where land adjacent to the park was incorporated into the design, and higher prices for the housing helped to offset the cost of construction. No doubt Olmsted gloated over the fact that property values adjacent to Central Park increased by more than $200 million over a seventeen-year span, from just before the park opened in 1859 until 1873. For Olmsted and the real estate speculators who benefited from the park, the increased property values more than compensated for the $13 million cost of constructing the now cherished urban greenspace. Nevertheless, the park's benefits were uneven. Poor property owners adjacent to the park unable to pay the higher assessments were forced to sell, while wealthy landowners saw the value of their properties increase dramatically (Rosenzweig and Blackmar 1992).

While Central Park and parks that followed helped to boost property values and thereby "commodify" nature, parks had other purposes as well. Pastoral elements of the large urban parks drew on nineteenth-century Romanticism. A reaction to the rationalism of the Enlightenment and the geometric and imperial park designs that accompanied it (as seen, for example, in Versailles), Romanticism celebrated feelings, inspiration, and creativity. For the Romantics, nature stood in opposition to the horrors of the industrial world. Parks, although highly contrived, represented the virtues of the natural world. For Olmsted and others, such places could have curative effects for mind, body, and soul. Among other things, the open space provided sunlight and clean air for city dwellers living in congested and polluted conditions. Parks became metaphorically the lungs of the city, the mechanism to breathe in the good and flush out the bad. Quite literally, in the age of miasmas, parks were promoted for improving public health.

Parks were also promoted as a way to soothe angry working-class mobs and aid the moral development of society. Social engineering, and the faith that "nature" could help accomplish it, was a strong incentive for Victorian parks. An idealized English countryside served as the archetypical urban pastoral scene. The construction of parks also had a concrete impact on social relations in the city. When the land for Central Park was acquired by eminent domain, it was already occupied by more than a thousand people living in shantytowns. Poor Irish, German, and black inhabitants were swept aside for the construction of Central Park. Despite the democratic principles of the park, initially it was a space only for those wealthy enough to afford a carriage that would take them to the park, which at the time was not "central" to the city. Not until cheaper mass transit became available could a broader spectrum of society enjoy the park (Rosenzweig and Blackmar 1992).

Parks in the United States took their cue from the large urban parks of western Europe. Hyde Park in London covers 350 acres in the borough of Westminster. Originally the grounds for Westminster Abby, the area was acquired by Henry VIII in 1536 and used for royal hunting parties. The park was not opened to the public until 1637 and then only on a limited basis. In the 1730s the artificial serpentine lake was constructed, a feature that became typical of North American parks in the nineteenth century. Hyde Park inspired Baron Haussmann, responsible for the monumental redesign of Paris in the 1850s, to construct the twenty-two-hundred-acre Bois de Boulogne on the outskirts of Paris. Unlike the formal, geometric Luxembourg Gardens near the center of Paris, the Bois de Boulogne was designed in a more organic fashion, with artificial lakes and ponds and meandering pathways through wooded areas. The park's design reflected the Romanticism of the period but also the hygienists' desire to construct the park as the "lungs" of the city (Harvey 2003). Hyde Park and the Bois de Boulogne influenced park designs as far away as South America. Rio de Janeiro's Campo de Santana, built between 1873 and 1880, was constructed as a miniature Bois de Boulogne. It was later embellished by the prefect Pereira Passos, during the grand reforms of the city, which were modeled after the Haussmann reforms of Paris, between 1902 and 1906 (Boone 1994; Boone 1995; Needell 1984). Like Central Park, the Campo de Santana was originally elite space. The idea that greenspace was elite space would be contested in the twentieth century.

PLAYGROUNDS AND PROGRESSIVES

By the 1890s the idea of the large pastoral park began to be challenged. The rapid expansion of cities that went hand in hand with industrializing

FIGURE 6.1. Children on swings in Central Park, New York City, 1871. In the decades after this image was produced, advocates pressed for smaller neighborhood parks and playgrounds scattered throughout cities instead of large parks on the periphery. The playground movement was based largely on the principle of park access to children from all socioeconomic classes, rather than just the finely dressed, as shown here.
Source: Reproduced by permission of the New York Public Library.

economies meant that large tracts of land were more expensive than they had been in the past. To enjoy the park, citizens often had to travel to it, which consumed time and money. As streetcars rumbled down thoroughfares, especially after electrification in the late 1880s, streets became far more dangerous places for children. Small neighborhood parks, or playgrounds, offered children a safer alternative to the increasingly congested streets (McShane 1988). At the same time, the increased mobility offered by the electric streetcar, and later the automobile, meant that city dwellers had improved access to the "real" countryside rather than the "constructed" countryside of Victorian parks. Reforms ushered in during the Progressive Era in the United States undermined the elitist elements of the large parks and called for play spaces and recreation areas that would serve a larger number of city dwellers.

The distinguishing feature of playgrounds was their size. Playgrounds were considerably smaller than large urban parks, usually

between one and ten acres, because they were meant to serve neighborhoods rather than entire cities. The high cost of land in built-up areas
also necessitated smaller plots of land. Playgrounds tended to be distributed throughout residential areas in order to provide access to as
many people as possible. Unlike the large Victorian parks, where
strolling and the passive appreciation of scenery were encouraged, playgrounds were designed for active play, with a special focus on children.
On playgrounds, equipment for games took precedence over landscaping or scenery (Tuason 1997). The incorporation of playgrounds into
schoolyards helped to reinforce and promote the idea of open space
where children could play and exercise. Over the course of the twentieth century, the National Recreation and Park Association in the United
States has been a strong and effective lobby for playgrounds of all types,
from simple grassy patches to skate parks, water parks, and even mini-
golf parks. Similar associations exist in other countries and have been
key proponents of preserving open space in cities and promoting playgrounds. Given their focus on children, communities have generally
been supportive of playground initiatives. In many cities they are important social gathering places, providing rare common spaces in urban
areas dominated by private space (Mitchell 2003).

CITY BEAUTIFUL AND MONUMENTAL GREENSPACES

Jacob Riis's famous photographs of the tenements of New York put into
stark focus what most people already knew—that hundreds of thousands of people lived in deplorable conditions in industrial cities.
Charles Dickens had illustrated the dehumanizing living conditions of
the urban poor in the workhouses of Britain's industrial cities, particularly in his novel *Hard Times* (1854). Engels and Marx drew on the experience of the working poor in the industrial cities of Britain for their new
vision of government and society (Engels 1987). Industrialization radically changed lives, removing increasing numbers of people from the
rhythms of agricultural life and subjecting them to the demands of
industrial production. The rapid growth of cities that accompanied
industrialization meant crowded conditions in inadequate housing, as
well as increasingly unsanitary conditions. All of this alarmed people
in power, partly out of pity but primarily out of fear (Boyer 1978). The
Paris Commune of 1858, the Chicago Haymarket riots of 1886, and the
Pullman Strike of 1894 struck fear into the hearts of city leaders and the
elite (Harvey 2003). The City Beautiful Movement was born in part as

a means of calming the seething masses, as a way to bring order and peace, as well as moral guidance, to the angry mobs.

The "White City" of Chicago's 1893 World's Fair was the ultimate expression of City Beautiful. Drawing on the École des Beaux Arts tradition in Paris, the monumental architecture, painted a gleaming white, stood in contrast to the dirt and filth of the nearby city of Chicago. Although it was an artificial city in many respects, sporting plaster façades and characterized by a complete absence of crime, it inspired awe in the millions of viewers who saw it. Visitors wanted their cities to look like the White City. City Beautiful's emphasis on grand buildings and civic squares influenced city planning for a generation (Wilson 1989).

Open space and parks were integral components of the City Beautiful Movement. The use of greenspace was a deliberate countermeasure to the crowded, squalid conditions of Chicago and other industrial cities. While parks created space between the celebrated buildings, they also reflected the monumentalism of the City Beautiful idea. Unlike the romanticized countryside parks of the Victorian era, the City Beautiful parks and open spaces tended to be rigid, linear, geometric shapes, often bounded by monuments or large civic buildings. The Mall of Washington, D.C., planned in 1901, is perhaps the best-known example of a City Beautiful greenspace. Nothing about the mall suggests an attempt to re-create the countryside in the city. Instead it stands as an imprint of imperial authority over space in the national capital. Ironically, the Mall has been used as a stage on which the "masses" have voiced their protests against government authority. From the civil rights speeches of Martin Luther King to the protests of the Iraq war, a space that had its origins in calming the masses has provided a space in which the masses can question and threaten authority.

GARDEN CITIES

In his spare time, an English stenographer named Ebenezer Howard tinkered with the idea of making cities into more humane environments. In contrast to the bleak industrial landscapes of the late nineteenth century, Howard envisioned a green city, neatly organized by specific land uses, leaving ample room for recreation, farming, and gardens. His prototypes for what he called the garden city show a defined center surrounded by concentric rings of housing, industry, and open land, all connected by radial thoroughfares and wide, tree-lined ring roads. Large areas of the city were dedicated to gardens, parks, and community allotments. Beyond the city, greenbelts of farmland would be kept in production and immune to the sprawl of large, metropolitan cities such as London. Howard envisioned a series of moderate-size cities—population no more than thirty thousand—connected by railway or canal to a central city of approximately sixty thousand, in other words a hub-and-spoke arrangement. In some ways Howard's ideal city is similar to the "edge cities" and multinucleated cities that have developed in the past fifty years, especially in North America, although the scale of modern cities, and certainly the principles, are very different from Howard's.

Howard's idea of the garden city falls in the category of utopian design. His ideal sprang from his political beliefs of cooperative socialism. Land in the garden city would be held in common, and the city, or cluster of cities, would be constructed away from the larger metropolitan areas where wealth and power were concentrated. For Howard, the garden city was more than just green space; it was an opportunity for a fresh start for community living removed from the power relations of established centers like London.

The plans were utopian on another level, in that his designs were set in open agricultural land, a blank slate for remaking the city unconstrained by existing urban morphologies. Rarely, of course, do visionaries get the opportunity for planning on such grandiose terms (Brasilia is an exception). Very often, most large-scale, top-down planning approaches are miserable failures. Howard recognized, however, that his ideal was utopian. On some of his sketches he made clear that the layout of the garden city would depend on the characteristics of the site (Fishman 1988). Unlike utopian novels of his time, however, Howard's plans were for nearby places rather than faraway mythical places and were meant to be real mechanisms for reform (Schuyler 2002). His designs were grounded in another way—through the establishment of garden city associations in nearly a dozen countries. Howard's revised *Garden Cities of To-morrow*, published in 1902, sold thousands of copies. After World War II it was republished with a foreword by Lewis Mumford, which rejuvenated interest in Howard's ideas (Parsons and Schuyler 2002).

The renowned urbanist Jane Jacobs has charged that Howard was partly responsible for encouraging suburban development through the decentralized principles embodied in his garden city philosophy (Jacobs 1961). Others have suggested that he was one of the first new urbanists, encouraging design on a human scale with access to open space and community ideals (Fishman 2002). New urbanism, in other words, may not be all that new (Fulton 2002).

SUBURBS AND PRIVATE GREENSPACE

Howard's ideas were manifested explicitly in a limited number of garden cities, most famously Letchworth and Welwyn in England. But his greatest impact on the design and function of cities can be found in the suburbs that ring urban cores with remarkable homogeneity. If Jane Jacobs is right, then Howard must surely be turning in his grave. His communitarian ideals of the garden city were co-opted by land subdividers and developers to create acre upon acre of *private* greenspace in the forms of yards, rather than the community spaces he envisioned. Even Letchworth, the first garden city, compromised Howard's ideals, as its board of directors created a middle- and upper-class enclave rather than a workingman's utopia (Ward 2002).

While the suburbs may be a far cry from Howard's garden city ideal, they continue nevertheless to be attractive places to live. In the United States between 1940 and 2000, the population of central cities has remained relatively stable, while suburbs have grown substantially. In 2000 half the country's population, or 140 million people, lived in suburbs. Nineteen-seventy marks the first year that more people lived in suburbs than in cities (Hobbs and Stoops 2002). The U.S. Census Bureau defines a suburb as the area inside a metropolitan area but outside the central city's boundaries. One could quibble over definitions, but the trends are clear to visitors to any large metropolitan area—the suburbs are growing faster than central cities and show no signs of slowing.

Geographers tend to talk about push and pull factors when describing reasons for migration, or why people choose to move from one place to another. Push factors are those that make a current location unattractive, while pull factors are qualities that make another location attractive. Moments of crisis often act as tipping points for migration (Jokisch and Pribilsky 2002). For many North Americans, fear of poor schools, crime, declining services, and high land or housing costs—whether perceived or real—are push factors from urban cores, while the opposite characteristics of suburbs are important pull factors. Race and class are also critical factors. White flight from American cities, especially after the Great Migration of southern blacks to northern cities, in addition to the race riots of the 1960s, remains a real factor in decisions to move. A house in the suburbs surrounded by a patch of green grass stands as a testament to middle-class success. The suburbs have a strong allure for middle-class Americans, Canadians, and Australians, and increasingly for other nationalities as well. It seems that the typical suburban layout of detached homes and private lawns is a winning formula (Lemon 1996).

But it is not always easy to separate what the market wants from what the market gets. Suburban homes continue to attract buyers, but developers have tended to build using a one-sizes-fits-all model. The veneer of custom-built façades in what are otherwise homogeneous environments is a shallow nod to individual expression in many suburban subdivisions. In a land of choice, residents may not have all the choices they want in where or how to live, or be informed of the alternatives. Buyers make tradeoffs in their decision to buy a house in the suburbs. They may trade the energy of city life for the quiet of the suburbs, or the lively streets of the city for less expensive housing out of town. If given the opportunity, buyers might be willing to give up a part of their lawn for community space, such as a park, or be willing to give up a few square feet in exchange for a convenient shopping district or nearby workplaces. Latent demand for urban services in suburban tracts is not always satisfied.

On the other hand, is there any point in trying to remake the city in the suburbs? Would doing so simply promote the continued outward expansion of cities onto farmland, the sprawl so often derided? An alternative approach might be to make the suburbs so horribly dull that no one wants to live there. Teenagers who endure a monolithic suburban landscape might prefer to stay away from them forever. Indeed, the gentrification of inner cities by the baby boom generation can be explained in part by a countercultural distaste for suburban life (Ley 1983).

GREENWAYS: IDEAL URBAN GREENSPACE?

Demand for land usually makes greenspace in central cities an expensive proposition. This is particularly the case if treeless, parkless centers have to "retrofit" with landscaping. Trees in planters or grass strips in medians can provide some limited benefits to humans and the environment and sometimes are the only practical solution for densely built-up areas of the city. Large, rectangular parks in central cores are difficult to establish if these areas already contain built structures. In the past twenty years many municipalities have studied the possibility of building linear parks, or greenways, as a practical alternative to traditional parks.

Deindustrialization and technological change have left many downtown areas with abandoned rail lines (as well as abandoned land, as discussed in the brownfields section below). Pulling up rails for scrap left gently graded paths that some recognized as perfect for cycling trails. The rails-to-trails movement began in the mid-1960s but picked up speed in the 1980s, especially after the Rails-to-Trails Conservancy was established in 1986. The rails-to-trails idea is an inventive reuse of abandoned land. In addition to the gentle grades, the rail trails are often separate from other traffic and are unbroken for long distances. Early efforts to separate dangerous railroad traffic from roadways have created cycling paths that are removed from much of the danger of cars and roads. In urban areas the rail trails often go to the heart of the city, where industry was once located. This makes the trails ideal for suburb-to-city commuting. Most trails are designed for multiple uses, including cycling, walking, inline skating, and running. Gentle grades and smooth surfaces also make them accessible to wheelchairs. Some permit horseback riding, while others are used in winter for cross-country skiing and snowmobiling. Where trails pass over bodies of water, people can fish off bridges. By September 2001 more than 100 million users traveled on twelve thousand miles of rail trails in the United States. The rail trail idea has spread all over the world, particularly to Europe, Canada, Mexico, Australia, and New Zealand (Fabos 1995; Rails-to-Trails Conservancy 2004).

Because many of the rail lines that get converted to trails have been abandoned for long periods of time, trees and other vegetation have had a chance to grow up alongside them. The trees provide shade, block wind and noise, and can serve as habitat for wildlife. In many ways the abandoned railways are like parks in that they provide green space for recreation and wildlife. For these reasons rail trails are sometimes called linear parks or greenways because of their distinctive shapes and their primary recreational purpose. Greenways have also been developed along rivers. These sites offer advantages similar to those of rail lines. Grades are gentle, and unless flood-control measures have been constructed are usually vegetated and undeveloped (Little 1990). Adding recreational trails is a way to secure the space for the public rather than for private development. In doing so, the environmental and social benefits of greenways are maintained. Buffering the stream with grass and trees improves water quality and in some circumstances can reduce flooding, since the vegetation and soils hold precipitation and release water slowly over time, reducing the chances of flash floods. The Friends of the Los Angeles River have used these arguments to restore the riparian vegetation of the L.A. River in some places and have argued for returning channelized portions of this highly urban river back to a more natural state (Gumprecht 1999). In Ottawa, Canada, the Rideau Canal remains channelized, but along much of its length it is flanked by trees, grass, and bicycle lanes. Boats use the canal for recreation in summer, while in winter skaters use it for recreation and even as a means of getting to work.

Greenways serve multiple purposes. The earliest greenways were constructed primarily for recreation and were typically paved with asphalt. Newer greenways tend to be constructed with multiple benefits in mind. Research has demonstrated that greenways can provide important habitat for wildlife and can act as wildlife corridors. Greenways have also been constructed for historic preservation, such as towpaths for canals or linking historic sites by paths for visitors to enjoy (Searns 1995). Using relatively little land, greenways can link existing parks and conservation areas, creating broader access to these important areas for people and wildlife (Arendt 2004). Given their linear structure, the amount of land used in greenways is relatively small, making them inexpensive to acquire. If they follow old transportation routes, municipalities often have rights-of-way, making the paths relatively easy to acquire through eminent domain.

Greenways have also been promoted for social benefits. One of the difficulties with traditional parks is that they tend to be created on the fringe of cities, where land is available, undeveloped, and relatively inexpensive. This provides access for people who live in the suburbs near the parks rather than those living in the densely built-up core of the city. Greenways can help ameliorate this environmental injustice by providing greenspace in the inner city and also create routes to traditional parks on the city's edge. Greenway planners also laud the ability of the trails to link communities and provide interaction in segregated cities through a common resource. But access to trails can be uneven. A study from Indianapolis shows that minority and poor populations have disproportionately high access to greenways (Lindsey et al. 2001). While it does not report on who uses the greenways, most research shows that the poor and minorities are less likely to use parks. One study that specifically examined greenways in Raleigh, North Carolina, found that the typical user tended to be a white female between the ages of sixteen and thirty-four, with no children, well educated, and with a higher-than-average income (Furuseth and Altman 1991). In other

FIGURE 6.2. Greenways, such as the Gywynns Falls Trail in Baltimore, Maryland, pictured here, have multiple ecological and social benefits that make them an increasingly attractive planning tool for cities.
Source: Photo by authors.

words, building a greenway through minority and poor communities does not necessarily mean that those groups will use it. But if planners pay attention to the needs of communities, more people from all backgrounds may use greenways more often. An extensive study from Chicago shows that people who live near a proposed greenway value cleanliness, nature, beauty, safety, and access. Appropriate development was also a key variable. In less developed areas, respondents wanted the greenway to be more natural, while in densely built-up areas, residents wanted the greenway to reflect the racial or ethnic heritage and architecture of the neighborhoods (Gobster and Westphal 2004). The Chicago case study demonstrates that bringing stakeholders into the planning process may improve sense of ownership and encourage more use of the amenity across the demographic and neighborhood spectrum.

Greenways have become a hot planning tool, but the idea is not entirely new. Tree-lined streets and boulevards of the City Beautiful Movement served similar purposes and offered some of the same advantages as modern greenways. Landscape architects in the late nineteenth century also understood the value of linear parks (Fabos 2004). Olmsted designed a number of park systems that linked existing parks with narrow green corridors, often following rivers and streams. His nephew, Frederick Law Olmsted Jr., proposed a plan for Baltimore in 1904 that included a series of stream valley parks and parkways to connect existing parks. It is noteworthy that Olmsted recognized that maintaining riparian vegetation would reduce flooding and erosion, and also serve as natural courses for storm water instead of building expensive underground conduits (Baltimore Department of Public Works 1926; Baltimore Municipal Arts Society 1904; Orser 2003).

While local governments and nonprofits struggle to create new greenway networks, many of Olmsted's greenways have been lost. The problem with Boston's "Emerald Necklace" of parks and greenways, according to Anne Whiston Spirn, is that it looks a lot like a beltway to modern planners. "Olmsted's emerald necklace of Boston, a model for modern greenways, slowly has been bisected, chopped up, filled in, by the needs of commuters. The linear park system unfortunately for park lovers resembled a modern beltway and the needs of the car and commuters would reign supreme" (Spirn 1984).

One wonders if the current greenway system might someday be threatened by development, much as Olmsted's park system was in Boston. As Jane Jacobs argues in her latest book (2004), the Western world seems to be suffering from amnesia, forgetting the important lessons from the past about creating livable, sustainable communities, a thought echoed by Dolores Hayden (2003). If we forget what happened to Olmsted's greenways, the same might happen to the current greenway networks, despite the enormous efforts government agencies and nonprofit groups have made to establish them. Conversion of community gardens to housing, a so-called higher use, is an example of how this can happen.

BROWNFIELDS: BENEFITS AND BARRIERS FOR URBAN REDEVELOPMENT

Another relic of the industrial era is the brownfield, defined as abandoned property that has real or perceived risks of contamination. Generally, brownfields are found near central cities in old industrial districts. Since many of the industries that once occupied the sites operated in an era with few or no environmental regulations, the soils may be contaminated with the toxic by-products of industrial production. Companies are generally wary of building on these sites for fear that they will expose employees to harmful substances or that they might be held liable for cleanup. The result is large holes in the urban fabric where potentially valuable land could be used to employ people, especially in inner cities, where unemployment levels are often higher than in suburbs. Offering employment in central cities has the added benefit of keeping residents in the city, thus reducing the potential for sprawl. Using a brownfield for redevelopment also means one less greenfield site, perhaps a farm, subjected to the bulldozer.

Given the multiple benefits, the U.S. EPA began a pilot program in 1995 to encourage brownfield redevelopment. There are approximately 450,000 brownfield sites in the United States, but the brownfield programs have rejuvenated only a tiny proportion of them. The EPA Brownfields Program makes funds available to inventory sites, for cleanup, and for environmental training. In 2001 Congress passed a bill to provide liability relief for small businesses that use brownfield sites and to grant funds to local agencies to assess, inventory, or clean up brownfield sites (EPA 2004). Limiting liability and providing funds for cleanup have been important to kick-starting brownfield redevelopment in the United States.

Success in redeveloping brownfield sites also depends on market conditions. Brownfield redevelopment is most likely to succeed if other factors, such as adequate infrastructure, are present. While federal and state

FIGURE 6.3. Brownfields are usually old, abandoned industrial sites that have real or perceived contamination. Redeveloping brownfields improves local conditions by removing hazardous products; by providing jobs in inner cities, where they are usually located; by using existing infrastructure; and by putting less pressure on greenfield sites for development. An eyesore and potential risk to nearby residents in a primarily African American community in Columbus, Ohio, this site is also an example of environmental injustice. *Source:* Photo by authors.

programs might encourage brownfield redevelopment, these programs alone are insufficient (Lange and McNeil 2004). In Toronto, a booming real estate market and a highly livable downtown has created demand for residential development on brownfield sites without government support. Industrial redevelopment of the sites, however, has not occurred to the same degree, leading more industry and employment to greenfield sites on the periphery. While residential development in central cities has multiple benefits, the migration of employment opportunities to the periphery can negate some of them. Although city planning has worked for more than a century to separate industry from residences, the Toronto study suggests that excluding industry from brownfield redevelopment has higher social, economic, and environmental costs than benefits (De Sousa 2002, 2003). Development of greenfield or suburban sites for industry increases infrastructure and service costs, reduces municipal tax bases, increases energy use and air pollution, reduces manufacturing employment in cities, and increases public health risks because contaminated sites are not cleaned up.

URBAN GROWTH BOUNDARIES AND GREENBELTS

In the 1960s Portland, Oregon, struggled with the same ills as most other American cities. Suburban sprawl onto rich agricultural land, growth of automobile use, and a declining downtown mirrored the changes being wrought in cities across the country. In an impassioned speech before the state legislature, Governor Tom McCall railed against urban sprawl and, in an extraordinary turn of events, managed to convince lawmakers to pass a sweeping bill requiring the planning and establishment of urban growth boundaries. Today Portland is celebrated as one of the most livable cities in the country, with a thriving downtown, compact form, high use of public transit and cycling, and productive farms on the rural-urban fringe. The growth boundary is not a permanent wall, but proponents have been very effective in limiting its expansion. Many attempts to remove the urban growth boundary have failed. The citizens of Portland have clung to the principle of the growth boundary with an almost religious enthusiasm, proclaiming the virtue of moderation, expressed in the boundary, versus the "vice of greed" personified by the free-market opponents to the regional planning scheme (Abbott 2002).

The growth boundary, however, is neither a new concept nor a North American one. In 1580 Queen Elizabeth issued a proclamation to establish a "cordon sanitaire" approximately five kilometers wide around London. This proclamation mandated that people "desist and forbear from any new building of any house or tenement within three miles from any gates of said city of London." However, Britain's greatest success at instituting a growth boundary came several centuries later with the advent of the greenbelt.

The genesis of the modern greenbelt can be traced to a number of influential thinkers in British planning, including Ebenezer Howard, Raymond Unwin, and Patrick Abercrombie. Abercrombie's *Greater London Plan 1944* contained proposals for a greenbelt up to sixteen kilometers wide around London that would curb urban growth, provide recreational possibilities, stimulate agriculture, and enhance the natural amenities of the area (Thomas 1970). Government policies adopted in the postwar period gave local authorities in Britain the power to control development and implement greenbelt plans, although the government's view of the purpose of greenbelts moved away from preserving amenity or providing recreation space. In 1961 the minister of housing and local government, Henry Brooke, explained the function of the greenbelt as "a stopper." "It may not all be very beautiful and it may not all be very green," he argued, "but without it the town would never stop, and that is the case for preserving the circles of land around the town" (Gault 1981).

The British public has viewed greenbelts as inviolable, but government policies prove that they are anything but. During the Thatcher years, policies "loosened the belt" to accommodate home building while paradoxically paying lip service to greenbelt land as sacrosanct. During the 1980s urban planners pointed to a new role for greenbelts—to assist in urban revitalization, particularly in regions of the country (e.g., the West Midlands) where deindustrialization had led to urban decline. Twenty years later the same arguments are being voiced in support of greenbelts, and the same battles over the development of greenbelt land for housing are being waged. Greenbelts now cover about 13 percent of England, although around .02 percent of greenbelt land in that country is lost every year to development (Stone-Lee 2004). In 2004 Scotland's government announced plans to review its greenbelt policy in light of pressures on the greenbelt from building developers (Denholm 2004). As the Campaign to Protect Rural England suggests, greenbelts are still working, but they remain under threat (Campaign to Protect Rural England 2001).

One criticism of greenbelts is that they can encourage "leapfrogging," or development beyond the greenbelt. Satellite cities are not necessarily a bad thing, and Howard incorporated such notions into his garden city plans. But if jobs do not leapfrog along with residences, then people may actually commute farther as a result. Greenbelts can become simply places to drive through rather than a bit of country on the edge of the city as they were originally intended. In Seoul, South Korea, the longstanding greenbelt, one scholar argues, has resulted in leapfrogging of development because people moved to the periphery faster than jobs did. Because more jobs are concentrated in the core, people have to commute beyond the greenbelt (Bae and Jun 2003). Jurisdictional issues can also complicate the effectiveness of greenbelts or growth boundaries. Although the urban growth boundary in Portland has been effective, spillover has occurred across the Willamette River into Washington State, where no such boundary exists (Jun 2004).

A second criticism of greenbelts as growth boundaries is that they increase real estate costs, making affordable housing scarce. While growth boundaries make cities livable, it is important to ask who benefits from the amenities of a compact city. A study of medium-size English cities found that higher urban densities led to better access to public transportation, improved access to services and facilities, and reduced levels of socioeconomic segregation, but decreased the availability of affordable housing and living space (Burton 2000). It seems reasonable to assume that limiting the supply of urban land would drive prices higher, but evidence from Portland suggests otherwise. One statistical study shows that Portland's housing prices have not risen as rapidly as those in other regions from 1980 to 2000, and that only in the first half of the 1990s did they exceed those of most other metropolitan areas. The author of this study argues that "it is erroneous to conclude from Portland's experience that UGBs [urban growth boundaries] inevitably cause home prices to rise faster" (Downs 2002, 7). Rapid increases in housing prices in the early 1990s have been attributed to an influx of Californians into Oregon and a speculative housing bubble, both of which had the effect of bringing housing prices in line with other metropolitan areas after depressed prices in the 1980s (Abbott 2002; Fregonese and Peterson 2002).

Even if urban growth boundaries increase property values, proponents argue that this is offset by the reduced need for cars and the heavy expenses associated with car ownership, as well as by lower taxes through more compact and efficient infrastructure. Since higher densities improve the efficiency of services such as public transit, fewer tax dollars are needed to support them when user fares cover more of the costs. Urban growth boundaries also mean the construction of less infrastructure and greater efficiency of existing infrastructure, which also reduces pressure on tax revenues. Supporters of the growth boundary also argue with the measurement of affordability associated only with single-family homes. They argue that residents are not just buying a house or renting an apartment; they are buying into a house and neighborhood bundled together. Even if the growth boundary increases housing prices, people are willing to pay for the benefits they receive from more cohesive, livable neighborhoods (Abbott 2002; Fregonese and Peterson 2002).

SMART GROWTH

Some people consider growth boundaries too restrictive and inflexible. As an alternative, "smart growth" uses government incentives to persuade developers to create more compact residential subdivisions. State programs use a bundle of incentives to discourage classic low-density sprawl. Some states restrict publicly financed infrastructure, such as sewers and water supply, only to developments that meet certain density standards. Grants for conservation easements are also given as part of smart growth plans. Other programs provide funds to encourage brownfield redevelopment, cash grants to homebuyers who live near their workplaces, or tax credits for employers who create jobs in existing central locations.

The state of Maryland is one of the earliest adopters of smart growth, and its attempts at curtailing sprawl have garnered national and international attention. In Maryland as in other states, land development and planning have traditionally been left in the hands of local governments. One of the consequences is a lack of coordination in planning, which has led to fragmented, ill-planned regional growth, especially on the fringes of metropolitan areas. Regional levels of government have been attempted in the United States in order to coordinate planning and improve efficiency of services, but these have been the exception rather than the rule (Wheeler 2002). Although cities and counties are creatures of the state, in general they have been left alone to plan, zone, or deliver their own services. The counties around Washington, D.C., and Baltimore have grown very rapidly over the past two decades, leading to classic suburban sprawl. In the 1980s, 210,000 acres of open space were bulldozed for development. The alarming growth rate prompted the formation of a Year 2020 Panel, which proposed that local government be responsible for planning, but with a state advisory board. The board was limited to an advisory position, so its recommendations lacked the teeth necessary to take a bite out of sprawl. In 1997 Maryland's governor, Parris Glendenning, spearheaded the Smart Growth Initiative, which backed up words with money.

Maryland's smart growth plan uses multiple methods to encourage compact growth, reduce commuting time, and redevelop central cities. At the heart of the program is the Priority Funding Areas Act, which limits state funding for infrastructure to existing communities and areas designated for smart growth. Priority funding areas include territory within the Washington and Baltimore beltways, enterprise zones, heritage areas, and other places designated by the state for revitalization. Counties can also designate priority funding areas if they meet residential densities and other land-use and service criteria. Among other criteria, newly planned communities must have a minimum density of 3.5 dwellings per acre to be designated a priority funding area (Haeuber 1999). The Rural Legacy Program is designed to protect rural areas from development by providing grants for conservation easements, greenbelts, or the purchase of development rights. The Job Creation Tax Credit provides state tax credits for companies that create jobs in central cities, while the Live Near Your Work Demonstration Program provides matching cash grants to home buyers who purchase homes near their place of work. The Smart Growth Initiative also sets aside funds for brownfield redevelopment to encourage the reuse of existing properties rather than the development of greenfield sites.

As one would expect, smart growth has come under attack from real estate lobbies that protest the restrictions and from environmentalists who find the regulations too weak. Maryland's new governor has chipped away at some of Glendenning's efforts, including the sale of state-owned land. Some have a more radical view of smart growth, which is to stop or reverse urban sprawl by any means. In December 2004 so called eco-terrorists were suspected of destroying twelve houses and damaging dozens of others in a dispute over development near a bog ecosystem in Maryland, south of Washington, D.C. (Partlow and Harris 2004). No evidence was found relating the fires to eco-terrorists (in fact, company employees and firefighters were indicted for starting the blaze), yet the suggestion of eco-terrorism points to how highly charged the issue of sprawl and sprawl control has become.

Nearly every governor in the United States has made a statement supporting smart growth, although their definitions and record of implementation have differed widely (Smart Growth Network 2004). Since no government leader wants to be seen as someone who promotes "dumb growth," smart growth, like sustainable development, has many jumping on the bandwagon, even if their actions do not match their words (Hayden 2003). Although many individuals enjoy the amenities of low-density suburbs, sprawl has negative connotations. Smart growth in a sense offers a way for people to have their cake and eat it too. It allows the continued growth of suburbs, but with a nod to sustainable development. Elected officials have supported the idea in part because it is popular. Despite the rhetoric, smart growth still means that growth will happen in the suburbs and on the periphery. In reality, the smartest growth, in terms of efficiency of services and environmental benefits, would be to encourage the rejuvenation of existing neighborhoods, taking advantage of existing infrastructure, preserving greenfield sites on the fringe, and reaping the energy benefits of renovating existing homes and neighborhoods rather than building new ones.

NEW URBANISM

Given the poor performance by government and developers in reducing sprawl, designers have sought to slow or reverse the trend. The most famous proponents of design-as-fix are the architects Andres Duany and Elizabeth Plater-Zyberk, partners in the firm of DuanyPlater-Zyberk and Company, better known as DPZ. Beginning with their influential plan for Seaside, Florida, in 1981, DPZ has designed hundreds of plans that incorporate neotraditional design with narrow streets, public spaces, and all types of conveniences for pedestrians and cyclists that encourage walking and bicycling rather than driving. Their projects incorporate the principle that the design of streets and pathways should take precedence over the design of buildings. Given the right street network, one with shaded sidewalks and narrow streets built on a grid, a good working city will follow. In Seaside, which ironically was the set (slightly modified) for the neo-Orwellian movie *The Truman Show,* modest-size traditional houses with porches and picket fences sit on small lots and face narrow streets. Space is allotted for pathways between buildings as well as paths to the beaches. Seaside's developer, Robert Davis, wanted to re-create the beach communities of his youth on land he inherited from his grandfather. DPZ drew inspiration from the wooden-frame vernacular cottages of Florida, constructed to adapt to the warm climate by incorporating large overhangs and windows for cross-ventilation. With eighty acres to work with, the site was deemed best for the development of a small town. Skeptics believed the development would be a money loser, but Seaside has been a resounding financial success, with rapidly rising property values. More important, it was a successful idea that has since been copied many times. Seaside and the other developments that followed offered an alternative to the shortcomings of the traditional suburb (Hayden 2003; Logan 2001).

Most suburban developments are easy targets for the new urbanists (Duany et al. 2000). Low housing density and curvilinear street patterns make infrastructure inefficient and public transit expensive, and promote driving for the simplest of errands. Visiting a neighbor without going through fenced backyards can mean a very long walk, or more likely a trip in the car. The very young and the old without access to cars must depend on parents or others for mobility. As the population ages, an increasing number of people are stranded in the suburbs. In the United States, one in five persons over the age of sixty-five does not drive, and more than half stay home on any given day because they have no transportation. By 2025 the number of Americans over sixty-five will increase by 80 percent to more than 62 million. Those living in sprawling suburbs will have the fewest options for alternatives to the private car (Bailey 2004). For the population that does drive, living in the suburbs means long daily commutes. DPZ has remarked on the sacrifices commuters make in order to live in the suburbs. While workers fought for the eight-hour day in the early twentieth century, today's average commute of one hour each way has returned the effective working day to ten hours. Two hours a day spent commuting is the equivalent of twelve workweeks. "In a society that provides its citizens only two to three weeks' annual vacation," writes DPZ, "that is a dismaying figure" (Duany et al. 2000).

Clearly, the suburbs were built around the technology of the car, which for the new urbanists and other urban critics is a failure. Wide streets and driveways (meaning fewer parked cars on the street) encourage higher speeds in areas that typically house a lot of young families with children. In some suburbs sidewalks are entirely absent or at best underused, since most destinations are longer than the average walk. Fear of mixed land use has resulted in zoning laws that keep residential areas separate from commercial and industrial areas. This means that most suburbanites have to get in the car if they need so much as a loaf of bread or a jug of milk. Apartments above shops, which reduce commutes to minutes instead of hours, are illegal in most suburban developments, despite the obvious social and environmental benefits that derive from such building arrangements. New urbanists and other also criticize the aesthetics of suburbs, remarking that they are "placeless" copies of one another, that they celebrate the individual over the neighborhood, and that they lack beauty (Relph 1987). James Kunstler, perhaps the most outspoken critic of modern architecture and urban design, has described the building of the past fifty years as "depressing, brutal, ugly, unhealthy, and spiritually degrading" (Kunstler 1993, 10). He takes aim particularly at building in the suburbs, which he describes with palpable disgust as

> jive-plastic commuter tract home wastelands, the Potemkin village shopping plazas with their vast parking lagoons, the Lego-block hotel complexes, the "gourmet mansardic" junk-food joints, the Orwellian office "parks" featuring buildings sheathed in the same reflective glass as the sunglasses worn by chain-gang guards, the particle-board garden

apartments rising up in every meadow and cornfield, the freeway loops around every big and little city with their clusters of discount merchandise marts, the whole destructive, wasteful, toxic, agoraphobia-inducing spectacle that politicians proudly called "growth." (Kunstler 1993, 10)

Kunstler continues to rail against modern architecture with a regular "Eyesore of the Month" web page (http://www.kunstler.com/index. html). Here he adroitly exposes the absurdities of modern building, urban design, and zoning practices. What is remarkable to Kunstler and others is that we seem to have learned little from our past mistakes and continue to build in the name of "growth" with little regard for quality of life (Hayden 2003; Jacobs 2004).

Although new urbanism has taken hold in North America, it is not without its critics. Many have argued that it is elitist, creating spaces only for the well-to-do. Others have suggested that new urbanism is steeped in nostalgia, harkening back to a past that never really was. Although Paul Goldberger, the architectural critic for *The New Yorker*, lauds much of what new urbanism has accomplished, he finds the physical determinism—the notion that good design will make good citizens—a weakness of the design philosophy. He also argues that new urbanism has a devoted group of followers who have an almost blind faith in the movement. He describes the annual meeting of the Congress for New Urbanism as "somewhere between a revival meeting and an academic conference" (Goldberger 2000, 128).

Many have noted that new urbanism is a misnomer, since most of its projects have occurred on greenfield sites. This has led some critics to dub it "new suburbanism." To be fair, DPZ argues that they encourage clients to look first in existing neighborhoods for the quality of life new urbanism promotes. DPZ has also undertaken some infill projects in cities (Duany et al. 2000). Goldberger suggests, however, that "the New Urbanists may never be fully comfortable with big cities because cities will never be wholly controllable, and control is what Duany and Plater-Zyberk want" (Goldberger 2000, 134). For Dolores Hayden (2003), new urbanism suffers from two shortcomings—the concentration on new development rather than existing "old-growth" suburbs, and the inability to address the larger political and economic structures that encourage sprawl. Clearly, design can play a role in mitigating car use, encouraging walking and cycling, and creating opportunities for interaction with neighbors. New urbanism has boldly challenged existing practices, but on its own it will not be enough to change the tide of growth on the fringe of cities.

PATHWAYS TO A SUSTAINABLE URBAN FUTURE

What are the pathways to a sustainable urban future? It is difficult to predict what lies ahead, but experience and knowledge can be the basis for guidelines or prescriptions. A precautionary approach, many would argue, is better than blind faith that all will be better, especially since the status quo cannot be sustained. Cities are immensely dynamic places, and they will change. After sixty years of cheap energy and subsidized suburbanization, the present generation is left with cities that may prohibit a healthy, high quality of life over the long term (Hayden 2003). As the world becomes urban, we have reached a defining moment. Now is the time to plan as wisely as possible for the future.

The first and most important prescription for a sustainable future is to be *smart*. This does not necessarily mean advocating "smart growth," although smart growth polices have merit. It suggests that we use our best ideas to create an urban future that improves quality of life while continuing to make cities vibrant, progressive, energetic, and inviting places. This ideal can be achieved if we learn to develop a consensus about the kind of cities we want to live in and leave to our children. Cities have been wonderful places in the past, and they can be wonderful in the future. We must also learn to be smart about the use of resources and the generation of waste. In part this depends on increasing efficiencies through improved technology, but the simplest method is to promote conservation in every way possible. Cities need to be designed and managed in ways that make conservation easy, as this is the least complicated and least costly way to benefit both the economy and the environment.

A second prescription is to build the city around the person rather than the car. Designing and constructing cities to accommodate the automobile has quite literally dehumanized the city. Driving can be a fun and liberating experience, but few people revel in traffic jams, or in having to grab the keys, start up the car, and drive a mile to pick up a loaf of bread. In the United States, commuting times have increased to nearly an hour per day, resulting in lost productivity and reduced quality of life. Behavior on the road is usually far worse than on the sidewalk; the segregation of people in fast-moving steel cages can bring out behavior rarely seen when people share a sidewalk or bike path. Millions of acres of concrete and asphalt for vehicle travel and storage could be used for greenspace, sidewalks, housing, and other purposes. Unfortunately, automobile use is growing worldwide. As we have seen, the growth in the number of motor vehicles on the road is higher in China than any-where else. Countries that follow the path of the United States and allow the automobile to dominate the urban landscape may be doomed to the same fate.

A third prescription is the careful redesign of existing cities and sub-urbs. Built forms are ultimately the product of human decisions. Human beings have the capacity like no other species to shape the environment to suit our needs and wishes. However, our decisions are limited by what we have to work with. The built form of cities represents a huge amount of inherited fixed capital. Urban dwellers are stuck with the products of past decisions, many of which were made with little con-cern for sustainability. Cities and especially suburbs have been designed in ways that make them difficult to retrofit or redesign into sustainable forms. Yet it is not an impossible task. Infill housing, apartments above shops, and mixed land use all work to increase densities and promote transportation choices other than the car. Careful redesign can remake urban landscapes on a human scale. The difficulty is that zoning restric-tions make such arrangements illegal or at the very least difficult. Redesign in this regard must be accompanied by a political willingness to allow more flexible zoning laws.

Increased commuting times, ironically, can promote more sustain-able urban forms by encouraging the development of edge cities in a multinucleated form. Jobs and services once found in the central busi-ness district can now be found in suburban clusters, especially near the intersections of major highways. While the car is still promoted as the usual transportation choice, other forms are possible with careful plan-ning. This is especially true when older, once separate small cities become engulfed in the suburban sprawl of larger cities. Many of these older towns are cleaned up and promoted for their historic value. Even if the main streets become outdoor shopping malls filled with national chains, people flock to them for their historic charm. While some may scoff at their inauthenticity, such places expose people to alternatives to typical suburbs. Other cities are reintroducing mass transit systems along old rights-of-way. Despite the enormous costs of retrofitting, Los Angeles has been working to rebuild its streetcar system. If it can hap-pen in Los Angeles, it can happen anywhere.

A fourth prescription for the sustainable city is to internalize exter-nalities. If people, businesses, and other organizations had to pay the true costs of energy use and waste disposal, behavior would almost cer-tainly change in ways that would promote sustainability. We amortize infrastructure construction so that the present generation does not pay all the cost; we should also amortize costs to the environment, which at present are treated as free. Environmental accounting is not an entirely satisfactory means of creating the sustainable city, for in the end it suggests that we can pay for or buy our way out of problems. Nevertheless, it can be an effective part of a strategy toward sustainability.

A fifth prescription for sustainability is to adopt a broader conception of health. To maintain human health we need to consume nutritious foods and clean water, exercise and breathe clean air, and avoid risks that might cause disease or death. People, at least those who can afford to, look after themselves today with an eye to enjoying a long, high-quality life. The same ideas can be applied to the city—as a place that is good for human and ecosystem health over the long term. The Healthy Cities Initiative, described in Chapter 5, takes a systematic approach to human well-being and the environment that mirrors new approaches to sustainability. Others have suggested the related concept of metabolism as metaphor for the dynamics of human and ecological systems. However you choose to look at it, if people begin to treat the health of the human ecosystem with at least the same respect and thoughtfulness they do (or should) their bodies, both human and ecosystem health would benefit greatly and lead to more sustainable ways of life in cities.

The sixth prescription for the sustainable city is the promotion of justice and equity. More than 2,300 years ago, Aristotle remarked that people come to cities, or remain there, to pursue opportunities for a better life, an observation that still rings true. In order for cities to provide sustained opportunities for the good life, those experiences must be just and fair. Few ideas survive very long that perpetuate human injustice. For cities to be sustainable they need to distribute the costs and benefits of urban living more equitably than they now do. The future of cities looks grim if societies continue to sweep problems under the rug, which in many cases means putting the worst by-products of cities in poor and minority neighborhoods. Yet there are hopeful signs on the horizon. The environmental justice movement, for example, has coincided with a significant reduction in toxic releases and overall improvement in air and water quality. The simple but profound decision to inform people about the risks of living in cities has been extremely effective, not only for communities that bear a disproportionate burden but for urban environments as a whole. Industry has also become more conscious of its externalities by having to report those figures to the

EPA. The environmental justice movement has taken hold in many parts of the world and is increasingly a force to be reckoned with.

The seventh prescription is to look to the past and elsewhere for examples of livable, human-scale cities. Too often we forget that people have struggled and thought a great deal about the same issues urban planners and visionaries tackle today, and yet we labor to re-create what has already been conceived and practiced (Hayden 2003). Critics suggest, for example, that so-called new urbanism is not really new at all, but a re-creation of building and design methods that worked well in former generations. Historical amnesia, Jane Jacobs argues, leads not only to redundancy but might also spell catastrophe for Western society (Jacobs 2004). If we do not take advantage of the past for its lessons, we follow a path of continued unsustainability. Arrogance is a dangerous and foolish thing.

North American cities are particularly prone to wasteful, unsustainable practices. Planners, urbanists, and individuals should draw on lessons from elsewhere when making decisions that affect their cities. European cities use far less energy per capita than North American cities and are known for their livability (Beatley 2000; Wheeler and Beatley 2004). When American tourists walk the streets of Paris or London, they might wonder why these cities are so much more inviting and beautiful than the ones back home. In part, European cities benefited from being built before the car was king, but other policies and points of view also contributed to making these places cater more to the needs of people than to the automobile. Within North America there are plenty of examples of cities that work better than most in working toward sustainability. New York and Montreal draw million of tourists every year, in part because they possess a unique sense of place and exemplify how wonderful cities can be. Even in those cities that have been dominated by the automobile era, some, like Toronto and Portland, Oregon, are inventive and imaginative enough to make livable, healthy cities (Portney 2003).

In his recent book on Paris and modernity, David Harvey makes the insightful comment that "no social order can change without the lineaments of the new already being present in the existing state of things" (Harvey 2003, 16). Modernity was not a distinct break from the past, as it is so often characterized. Existing conditions allowed it to happen. Likewise, conditions today are ripe for a transition to sustainability as a guiding principle for policy and decision making. The modern environmental movement is more than thirty years old, and it has shaped the way people see themselves in relation to the natural world. The vast majority of Americans call themselves environmentalists, and environmentalism has a strong following around the world. Clearly, a shift in the way we think about environment and environmentalism is necessary for success. Focusing on the environment as separate from labor, economy, or culture might very well lead to the "death of environmentalism," as a recent influential essay argues (Schellenberger and Nordhaus 2005). Environmentalism must move beyond technical fixes and single issues if it is going to be a guiding political and moral force. Sustainability offers a conceptual framework that can breathe life into an integrated human-ecological environmentalism.

Sustainability is in everyone's interest, as it ensures a long, healthy, high-quality life for our generation and those to follow. While technology and policy may shift sustainability in different directions, ultimately it is the idea and philosophy of sustainability that must take hold. Ideas can change the world. While deciding what kind of accounting system, land-use planning, or transportation technology is important, the most critical step is a widespread acceptance of the principles of sustainability, a commitment to long-term thinking, and responsibility to future generations that integrates social and ecological concerns.

References

Abbott, Carl. 2002. Planning a sustainable city: The promise and performance of Portland's urban growth boundary. In *Urban sprawl: causes, consequences, and policy responses*, ed. Gregory D. Squires, 207–36. Washington. D.C.: Urban Institute Press.

Abu-Lughod, Janet L. 1993. The Islamic city: Historic myth, Islamic essence, and contemporary relevance. In *Urban development in the Muslim world*, ed. Hooshang Amirahmadi and Salah S. El-Shakhs. New Brunswick, N.J.: Center for Urban Policy Research.

Adler, Sy. 1991. The transformation of the Pacific electric railway—Bradford-Snell, Roger Rabbit, and the politics of transportation in Los Angeles. *Urban Affairs Review* 27: 51–86.

Agyeman, Julian, Robert D. Bullard, and Bob Evans. 2003. *Just sustainabilities: Development in an unequal world*. Cambridge: MIT Press.

Al Sayyad, Nezar, ed. 2001. *Hybrid urbanism: On the identity discourse and the built environment*. Westport, Conn.: Praeger.

American Farmland Trust. N.d. American Farmland Trust: Frequently asked questions. http://www.farmland.org/FAQ.htm#losttodevelopment (accessed 5 May 2004).

Amirahmadi, Hooshang, and Salah S. El-Shakhs, eds. 1993. *Urban development in the Muslim world*. New Brunswick, N.J.: Center for Urban Policy Research.

Anderson, James C. 1997. *Roman architecture and society*. Baltimore: Johns Hopkins University Press.

Arendt, Randall. 2004. Linked landscapes: Creating greenway corridors through conservation subdivision design strategies in the northeastern and central United States. *Landscape and Urban Planning* 68: 241–69.

Armstrong, Christopher, and H. V. Nelles. 1977. *The revenge of the Methodist Bicycle Company: Sunday streetcars and municipal reform in Toronto, 1888–1897*. Toronto: P. Martin Associates.

———. 1986. *Monopoly's moment: The organization and regulation of Canadian utilities, 1830–1930*. Philadelphia: Temple University Press.

Ayres, R. U. 2000. Commentary on the utility of the ecological footprint concept. *Ecological Economics* 32: 347–49.

Bacon, Edmund N. 1974. *Design of cities*. London: Thames and Hudson.

Bae, C.H.C., and M. J. Jun. 2003. Counterfactual planning: What if there had been no greenbelt in Seoul? *Journal of Planning Education and Research* 22: 374–83.

Bailey, Linda. 2004. *Aging Americans: Stranded without options*. Surface Transportation Policy Project. http://www.transact.org/library/reports_html/seniors/aging.pdf.

Baltimore Department of Public Works. 1926. Report and recommendation on park extensions for Baltimore. Baltimore: City Plan Committee of the Department of Public Works.

Baltimore Municipal Arts Society. 1904. *Report upon the development of public grounds for greater Baltimore.* Baltimore: Lord Baltimore Press.

Barles, Sabine. 2002. L'invention des eaux usées: L'assainissement de Paris, de la fin de l'Acien Régime à la seconde guerre mondiale. In *La pollution dans les sociétés urbaines et industrielles d'Europe,* ed. C. Bernhadt and G. Massard-Guilbaud, 129–56. Clermont Feraud: UBP.

Barlow, P. M., and E. C. Wild. 2002. Bibliography on the occurrence and intrusion of saltwater in aquifers along the Atlantic coast of the United States. United States Geological Survey Open-File Report 02-235.

Beatley, Timothy. 2000. *Green urbanism: Learning from European cities.* Washington, D.C.: Island Press.

Beder, Sharon. 1993. From sewage farms to septic tanks: Trials and tribulations in Sydney. *Journal of the Royal Australian Historical Society* 79: 72–95.

Bennett, M., and M. Newborough. 2001. Auditing energy use in cities. *Energy Policy* 29: 125–34.

Bhatnagar, S., U. Dosajh, and S. D. Kapoor. 1988. Correlates of morbidity and patterns of mortality in urban slums of Delhi: Part I. *Health and Population: Perspectives and Issues* 11: 83–95.

Blumenthal, U. J., D. D. Mara, A. Peasy, G. Ruiz-Palacios, and R. Stott. 2001. Reducing the health risks of using wastewater in agriculture: Recommended changes to WHO guidelines. *Urban Agriculture Magazine* 3: 26–29.

Bonine, Michael E. 1979. The morphogenesis of Iranian cities. *Annals of the Association of American Geographers* 69: 208–24.

———, ed. 1997. *Population, poverty, and politics in Middle East cities.* Gainesville: University Press of Florida.

Boone, Christopher G. 1994. The Rio de Janeiro Tramway, Light and Power Company and the "modernization" of Rio de Janeiro during the old republic. Ph.D. diss., Department of Geography, University of Toronto.

———. 1995. Streetcars and politics in Rio de Janeiro: Private enterprise versus municipal-government in the provision of mass-transit, 1903–1920. *Journal of Latin American Studies* 27: 343–65.

———. 1996. Language politics and flood control in nineteenth-century Montreal. *Environmental History* 1, no. 3: 70–85.

———. 2002. Obstacles to infrastructure provision: The struggle to build comprehensive sewer works in Baltimore. *Historical Geography* 31: 151–68.

Boone, Christopher G., and Ali Modarres. 1999. Creating a toxic neighborhood in Los Angeles county—A historical examination of environmental inequity. *Urban Affairs Review* 35: 163–87.

Botero, D., and J. I. Wolfsdorf. 2005. Diabetes mellitus in children and adolescents. *Archives of Medical Research* 36: 281–90.

Bouvier-Daclon, N., and G. Senecal. 2001. The Montreal Community Gardens: A socially ambiguous territory. *Loisir et Société—Society and Leisure* 24: 507–31.

Bowen, W. M. 1999. Comments on "Every breath you take . . .": The demographics of toxic air releases in southern California. *Economic Development Quarterly* 13: 124–34.

Boyer, Paul S. 1978. *Urban masses and moral order in America, 1820–1920.* Cambridge: Harvard University Press.

Briffett, Clive, Jeffery Philip Obbard, and Jamie Mackee. 2003. Towards SEA for the developing nations of Asia. *Environmental Impact Assessment Review* 23: 171–96.

Brockerhoff, Martin P. 2000. An urbanizing world. *Population Bulletin* 55: 1–44.

Brown, Allison. 2002. Farmers' market research, 1940–2000: An inventory and review. *American Journal of Alternative Agriculture* 17: 167–76.

Brown, Cheryl. 2003. Consumers' preferences for locally produced food: A study in southeast Missouri. *American Journal of Alternative Agriculture* 18: 213–24.

Brown, P. 1995. Race, class, and environmental health: A review and systematization of the literature. *Environmental Research* 69: 15–30.

Bryld, Erik. 2003. Potentials, problems, and policy implications for urban agriculture in developing countries. *Agriculture and Human Values* 20: 79–86.

Buckley, Geoffrey L. 2004. *Extracting Appalachia: Images of the Consolidation Coal Company, 1910–1945.* Athens: Ohio University Press.

Bullard, R. D. 1994a. Overcoming racism—Commentary. *Environment* 36: 3–4.

———. 1994b. Overcoming racism in environmental decision-making. *Environment* 36: 10.

Bullard, R. D., and G. S. Johnson. 2000. Environmental justice: Grassroots activism and its impact on public policy decision making. *Journal of Social Issues* 56: 555–78.

Bunce, Michael F. 1994. *The countryside ideal: Anglo-American images of landscape.* London: Routledge.

———. 1998. Thirty years of farmland preservation in North America: Discourses and ideologies of a movement. *Journal of Rural Studies* 14: 233–47.

Bureau of Transportation Statistics. 2001. The 2001 national household travel survey. http://www.bts.gov/publications/national_household_travel_survey/highlights_of_the_2001_national_household_travel_survey/ (accessed 21 July 2004).

Burney, N. A. 1995. Socioeconomic development and electricity consumption: A cross-country analysis using the random coefficient method. *Fuel and Energy Abstracts* 36: 436.

Burton, Elizabeth. 2000. The compact city: Just or just compact? A preliminary study. *Urban Studies* 37: 1969–2001.

Burton, S. 1999. Evaluation of healthy city projects: Stakeholder analysis of two projects in Bangladesh. *Environment and Urbanization* 11: 41–52.

California Air Resources Board. 2002. Zero-emission vehicle outreach materials. http://www.arb.ca.gov/msprog/zevprog/factsheets/factsheets.htm (accessed 25 February 2003).

Campaign to Protect Rural England. 2001. Green belts still working, still under threat. http://www.cpre.org.uk/new-releases/news-rel-2001/024a.htm (accessed 8 December 2004).

Carneiro, Robert L. 1970. A theory of the origin of the state. *Science* 169: 733–38.

Casico, Elio Lo. 1994. The size of the Roman population: Beloch and the meaning of the Augustan census figures. *Journal of Roman Studies* 84: 23–40.

Centers for Disease Control and Prevention, and National Center for Health Statistics. 2001. *Health, United States, 2001, with urban and rural health checkbook.* Hyattsville, Md.: National Center for Health Statistics. http://www.cdc.gov/nchs/data/hus/hus01.pdf.

Cervero, R., and M. Duncan. 2003. Walking, bicycling, and urban landscapes: Evidence from the San Francisco Bay area. *American Journal of Public Health* 93: 1478–83.

Cervero, R., and K. Kockelman. 1997. Travel demand and the 3Ds: Density, diversity, and design. *Transportation Research Part D—Transport and Environment* 2: 199–219.

Chakrabarti, P. G. Dhar. 2001. Urban crisis in India: New initiatives for sustainable cities. *Development in Practice* 11: 260–72.

Chomsky, Noam. 1993. *Year 501: The conquest continues.* Boston: South End Press.

City Farmer. 1997. Report on community and allotment gardening in the greater Vancouver region. http://www.cityfarmer.org/normgardens.html (accessed 12 December 2003).

Clark, John G. 1995. Economic development vs. sustainable societies: Reflection on the players in a crucial contest. *Annual Review of Ecology and Systematics* 26: 225–48.

CNNMoney.com. 2006. Gas prices around the world: Think you pay a lot for gas? Perhaps you'd prefer to live in Venezuela. http://money.cnn.com/pf/features/lists/global_gasprices/ (accessed 21 January 2006).

Cohen, Susan A. 1993. The road from Rio to Cairo: Toward a common agenda. *International Family Planning Perspective* 19: 61–66.

Colls, Jeremy. 2002. *Air pollution.* London: Spoon Press.

Colten, Craig E. 2002. Basin Street blues: Drainage and environmental equity in New Orleans, 1890–1930. *Journal of Historical Geography* 28: 237–57.

———. 2005. *An unnatural metropolis: Wresting New Orleans from nature.* Baton Rouge: Louisiana State University Press.

Condran, G., and E. Crimmins-Gardner. 1978. Public health measures and mortality in U.S. cities in the late nineteenth century. *Human Ecology* 6: 27–54.

Conway, Gordon. 1998. *The doubly green revolution: Food for all in the twenty-first century.* Ithaca, N.Y.: Comstock.

———. 2000. Food for all in the 21st century. *Environment* 42: 9–18.

Conzen, M.R.G. 1960. Alnwick, Northumberland: A study in town-plan analysis. *Transactions and Papers (Institute of British Geographers)* 27: iii, ix, xi, 1, 3–122.

Copenhagen Health Administration. 2004. Copenhagen healthy city. http://www.sundby.kk.dk/sundikbh/pegasus.nsf/url/inenglish (accessed 20 July 2004).

Craddock, Susan. 2000. *City of plagues: Disease, poverty, and deviance in San Francisco.* Minneapolis: University of Minnesota Press.

Crawford, Harriet E. W. 2004. *Sumer and the Sumerians.* Cambridge: Cambridge University Press. (1st ed. 1999.)

Crompton, John L. 2001. The impact of parks on property values: A review of the empirical evidence. *Journal of Leisure Research* 33: 1–31.

Cronon, William. 1991. *Nature's metropolis: Chicago and the great West.* New York: W. W. Norton.

Cruz, Maria Caridad, and Roberto Sánchez Medina. 2003. *Agriculture in the city: A key to sustainability in Havana, Cuba.* Kingston, Jamaica: Ian Randle Publishers/IDRC.

Cummins, Susan Kay, and Richard Joseph Jackson. 2001. The built environment and children's health. *Pediatric Clinics of North America* 48: 1241–52.

Davies, W. Paul. 2003. An historical perspective from the green revolution to the gene revolution. *Nutrition Reviews* 61: 124–37.

Davis, Mike. 1992. *City of quartz: Excavating the future in Los Angeles.* New York: Vintage Books.

Davis, Phillip. 2003. Water-filtration plant sparks concern about impact on ecology. National Public Radio Broadcast, 30 January 2003. http://discover.npr.org /features/feature.jhtml?wfId=949664 (accessed 24 July 2003).

De Sousa, C. A. 2002. Measuring the public costs and benefits of brownfield versus greenfield development in the Greater Toronto area. *Environment and Planning B—Planning and Design* 29: 251–80.

———. 2003. Turning brownfields into green space in the City of Toronto. *Landscape and Urban Planning* 62: 181–98.

DeJong, Jocelyn. 2000. The role and limitations of the Cairo international conference on population development. *Social Science and Medicine* 51: 941–53.

Delegates of the United Nations Conference on Environment and Development. 1992. The "earth summit" on population. *Population and Development Review* 18: 571–82.

Demographic and Health Surveys. 2004. STATcompiler. http://www.measuredhs. com/ (accessed 14 July 2004).

Denholm, Andrew. 2004. Green-belt rules to be overhauled as pressure to build rises. *The Scotsman,* 12 August 2004. http://news.scotsman.com/politics. cfm?id=926232004 (accessed 13 January 2006).

Diakonoff, I. M. 1969. *Ancient Mesopotamia: A socio-economic history.* Moscow: Nauka Publishing.

Dickinson, Robert E. 1945. The morphology of the medieval German town. *Geographical Review* 35: 74–97.

Dockery, D. W., C. A. Pope III, X. Xiping, J. D. Spengler, J. H. Ware, M. E. Fay, B. G. Ferris Jr., and F. E. Speizer. 1993. An association between air pollution and mortality in six U.S. cities. *New England Journal of Medicine* 329: 1753–59.

Donohoe, Martin. 2003. Causes and health consequences of environmental degradation and social injustice. *Social Science and Medicine* 56: 573–87.

Downs, Anthony. 2002. Have housing prices risen faster in Portland than elsewhere? *Housing Policy Debate* 13: 7–31.

Duany, Andres, Elizabeth Plater-Zyberk, and Jeff Speck. 2000. *Suburban nation: The rise of sprawl and the decline of the American dream.* New York: North Point Press.

Dublin Healthy Cities. 2004. Environment. http://www.dublinhealthycities.ie/ pages/environment.htm (accessed 20 July 2004).

Duhl, L. J., and A. K. Sanchez. 1999. Healthy cities and the planning process: A background document on links between health and urban planning. Copenhagen: World Health Organization Regional Office for Europe.

Dwyer, John F., David J. Nowak, Mary Heather Noble, and Susan M. Sisinni. 2000. *Connecting people with ecosystems in the 21st century: An assessment of our nation's urban forests.* Portland, Ore.: U.S. Dept. of Agriculture, Forest Service, Pacific Northwest Research Station.

Earle, Carville. 1990. Regional economic development west of the Appalachians, 1815–1860. In *North America: The historical geography of a changing continent,* ed. Robert D. Mitchell and Paul A. Groves, 172–97. Savage, Md.: Rowman and Littlefield.

The Economist. 2002. 5 September. http://www.economist.com/science/display Story.cfm?story_id=1313810

Ehrlich, Paul R. 1968. *The population bomb.* New York: Ballantine Books.

Elsheshtawy, Yasser. 2004. *Planning Middle Eastern cities: An urban kaleidoscope in a globalizing world.* London: Routledge.

Engels, Friedrich. 1987. *The condition of the working class in England.* London: Penguin Books. (Orig. pub. 1845.)

Environmental Protection Agency. 1996. Overview of the storm water program. Washington, D.C.: U.S. Environmental Protection Agency, Office of Water.

———. 2004. EPA Brownfields homepage. http://www.epa.gov/swerosps/bf/index.html (accessed 30 November 2004).

Eskeland, Gunnar S., and Tarhan Feyzioglu. 1997. Rationing can backfire: The "day without a car" in Mexico City. *World Bank Economic Review* 11: 383–408.

Evenson, Robert E., and D. Gollin. 2003. Assessing the impact of the green revolution, 1960 to 2000. *Science* 300: 758–62.

Ewing, R., Richard A. Schieber, and Charles V. Zegeer. 2003a. Urban sprawl as a risk factor in motor vehicle occupant and pedestrian fatalities. *American Journal of Public Health* 93: 1541–45.

Ewing, R., T. Schmid, R. Killingsworth, A. Zlot, and S. Raudenbush. 2003b. Relationship between urban sprawl and physical activity, obesity, and morbidity. *American Journal of Health Promotion* 18: 47–57.

Fabos, J. G. 1995. Introduction and overview—The greenway movement, uses and potentials of greenways. *Landscape and Urban Planning* 33: 1–13.

———. 2004. Greenway planning in the United States: Its origins and recent case studies. *Landscape and Urban Planning* 68: 321–42.

2002. Farming goes to town. *New Scientist* 174: 41.

Feather, Peter, and Charles H. Barnard. 2003. Retaining open space with purchasable development rights programs. *Review of Agricultural Economics* 25: 369–84.

Federal Transit Administration. 2000. Safety and security statistics and analysis. http://transit-safety.volpe.dot.gov/Data/Samis.asp (accessed 29 July 2006).

Few, Roger, Trudy Harpham, and Sarah Atkinson. 2003. Urban primary health care in Africa: A comparative analysis of city-wide public sector projects in Lusaka and Dar es Salaam. *Health and Place* 91: 45–53.

Fincher, Ruth. 1998. Population questions for Australian cities: Reframing our narratives. *Australian Geographer* 29: 31–47.

Fishman, Robert. 1988. *Urban utopias in the twentieth century: Ebenezer Howard, Frank Lloyd Wright, and Le Corbusier.* Cambridge: MIT Press.

———. 2002. The bounded city. In *From garden city to green city: The legacy of Ebenezer Howard,* ed. Kermit C. Parsons and David Schuyler, 58–66. Baltimore: Johns Hopkins University Press.

Florida Department of Transportation. 2005. *2004 transportation costs.* Florida Office of Planning Policy.

Food and Agriculture Organization (FAO). 2001. Urban and peri-urban agriculture: A brief guide for the successful implementation of urban and peri-urban agriculture in developing countries and countries of transition. Rome: Food and Agriculture Organization. http://www.fao.org/fcit/docs/briefing-guide.pdf (accessed 13 January 2006).

———. 2004. *The state of food and agriculture, 2003–2004.* Rome: Food and Agriculture Organization of the United Nations.

Foster, John Bellamy. 2000. *Marx's ecology: Materialism and nature.* New York: Monthly Review Press.

Fregonese, John, and Lynn Peterson. 2002. Costs of "dumb growth." *Architecture Week,* 8 May 2002, 21–22.

Friedman, M. S., K. E. Powell, L. Hutwagner, L. M. Graham, and W. G. Teague. 2001. Impact of changes in transportation and commuting behaviors during the 1996 summer Olympic Games in Atlanta on air quality and childhood asthma. *JAMA—Journal of the American Medical Association* 285: 897–905.

Frumkin, Howard, Lawrence Frank, and Richard Jackson. 2004. *Urban sprawl and public health: Designing, planning, and building for healthy communities.* Washington, D.C.: Island Press.

Fulton, William. 2002. The garden suburb and the new urbanism. In *From garden city to green city: The legacy of Ebenezer Howard,* ed. Kermit C. Parsons and David Schuyler, 159–170. Baltimore: Johns Hopkins University Press.

Fung, A., and D. O'Rourke. 2000. Reinventing environmental regulation from the grassroots up: Explaining and expanding the success of the toxics release inventory. *Environmental Management* 25: 115–27.

Furuseth, Owen J., and Robert E. Altman. 1991. Who's on the greenway? Socioeconomic, demographic, and locational characteristics of greenway users. *Environmental Management* 15: 329–36.

Gandy, Matthew. 2002. *Concrete and clay: Reworking nature in New York City.* Cambridge: MIT Press.

Gault, I. 1981. Green belt policies in development plans. Working Paper 41. Oxford: Department of Town Planning, Oxford Polytechnic.

Gelt, Joe. 1997. Constructed wetlands: Using human ingenuity, natural processes to treat water, build habitat. *Arroyo* 9: 4. http://ag.arizona.edu/AZWATER/arroyo/094wet.html (accessed 13 January 2006).

Geronimus, Arline T. 2001. Urban life and health. In *International Encyclopedia of the Social and Behavioral Sciences,* ed. N. J. Smelser and P. B. Baltes, 16026–29. Amsterdam: Elsevier.

Giddings, B., B. Hopwood, and G. O'Brien. 2002. Environment, economy and society: Fitting them together into sustainable development. *Sustainable Development* 10: 187–96.

Gladwell, Malcolm. 2004. The terrazzo jungle. *New Yorker*, 15 March, 120–27.

Gobster, Paul H., and Lynne M. Westphal. 2004. The human dimensions of urban greenways: Planning for recreation and related experiences. *Landscape and Urban Planning* 68: 147–65.

Goldberger, Paul. 2000. It takes a village: The anti-sprawl doctors make a manifesto. *New Yorker*, 27 March, 128–34.

Gottlieb, Robert, and Andrew Fisher. 1995. Community food security: Policies for a more sustainable food system in the context of the 1995 Farm Bill and beyond. Los Angeles: Ralph and Goldy Lewis Center for Regional Policy Studies.

Gould, W.T.S. 1998. African mortality and the new "urban penalty." *Health and Place* 4: 171–81.

Graunt, John. 1662. *Natural and political observations mentioned in the following index and made upon the bills of mortality*. London: Roycroft, for John Martin, James Allestry, and Dicas.

Greene, Richard P., and John Stager. 2001. Rangeland to cropland conversions as replacement land for prime farmland lost to urban development. *Social Science Journal* 38: 543–55.

Greenwald, M. J. 2003. The road less traveled—New urbanist inducements to travel mode substitution for nonwork trips. *Journal of Planning Education and Research* 23: 39–57.

Groffman, P. M., N. L. Law, K. T. Belt, L. E. Band, and G. T. Fisher. 2004. Nitrogen fluxes and retention in urban watershed ecosystems. *Ecosystems* 7: 393–403.

Großman, Katrin. 2004. Declining cities—Rising futures? Paper presented at the "City Futures" conference, Chicago, July 2004. http://www.uic.edu/cuppa/cityfutures/papers/webpages/cityfuturespapers/session2_6/2_6d ecliningcities.pdf.

Grove, J. M. 1996. The relationship between patterns and processes of social stratification and vegetation of an urban-rural watershed. Ph.D. diss., School of Forestry and Environmental Studies, Yale University.

Grove, J. M., M. L. Cadenasso, W. R. Burch Jr., S.T.A. Pickett, J.P.M. O'Neil-Dunne, K. Shwarz, M. Wilson, A. R. Troy, and C. G. Boone. 2006. Data and methods comparing social structure and vegetation structure of urban neighborhoods in Baltimore, Maryland. *Society and Natural Resources* 19, no. 2: 117–36.

Guillerme, Andrâe. 1988. *The age of water: The urban environment in the north of France, a.d. 300–1800*. College Station: Texas A&M University Press.

Gumprecht, Blake. 1999. *The Los Angeles River: Its life, death, and possible rebirth*. Baltimore: Johns Hopkins University Press.

Haeuber, Richard. 1999. Spawl tales: Maryland's Smart Growth Initiative and the evolution of growth management. *Urban Ecosystems* 3: 131–47.

Haines, Michael R., and Richard H. Steckel, eds. 2000. *A population history of North America*. Cambridge: Cambridge University Press.

Hancock, T. 2000. Healthy communities must also be sustainable communities. *Public Health Reports* 115: 151–56.

———. 2001. People, partnerships, and human progress: Building community capital. *Health Promotion International* 16: 275–80.

Hancock, T., R. Labonte, and R. Edwards. 1999. Indicators that count! Measuring population health at the community level. *Canadian Journal of Public Health—Revue Canadienne de Santé Publique* 90: S22–26.

Hanson, Susan, and I. Johnston. 1985. Gender differences in worktrip length. *Urban Geography* 6: 193–219.

Harvey, David. 2003. *Paris, capital of modernity.* New York: Routledge.

Haub, Carl. 2002. How many people have ever lived on earth? *Population Today* 30: 4–5.

Hayden, Dolores. 2003. *Building suburbia: Green fields and urban growth, 1820–2000.* New York: Pantheon Books.

Henn, Patrick. 2000. User benefits of urban agriculture in Havana, Cuba: An application of the contingent valuation method. Master's thesis, Department of Agricultural Economics, McGill University.

Herendeen, R. A. 2000. Ecological footprint is a vivid indicator of indirect effects. *Ecological Economics* 32: 357–58.

Hinman, Sarah E. 2002. Urbanization and public health: A study of the spatial distribution of infant mortality in Baltimore, Maryland, 1880. Master's thesis, Department of Geography, Ohio University.

Hobbs, Frank, and Nicole Stoops. 2002. *Demographic trends in the 20th century: Census 2000 special reports.* Washington, D.C.: U.S. Census Bureau.

Hoek, G., B. Brunekreef, S. Goldbohm, P. Fischer, and P. A. van den Brandt. 2002. Association between mortality and indicators of traffic-related air pollution in the Netherlands: A cohort study. *Lancet* 360: 1203–9.

Hornby, William Frederic, and Melvyn Jones. 1993. *An introduction to population geography.* Cambridge: Cambridge University Press.

Horvath, A. and C. T. Hendrickson. 1998. Comparison of environmental implications of asphalt and steel-reinforced pavements. Transportation Research Record 1626. Washington, D.C.: Transportation Research Board.

International Development Research Centre (IDRC). 2003. Overview of cities feeding people programming. http://network.idrc.ca/en/ev-4675-201-1-DO_TOPIC.html (accessed 8 December 2003).

Jackson, John Brinckerhoff. 1984. *Discovering the vernacular landscape.* New Haven: Yale University Press.

Jackson, Kenneth T. 1985. *Crabgrass frontier: The suburbanization of the United States.* New York: Oxford University Press.

Jacobs, Jane. 1961. *The death and life of great American cities.* New York: Random House.

———. 2004. *Dark age ahead.* New York: Random House.

Japan Highway Public Corporation. 2002. Annual report 2002. http://www.jhnet.go.jp/english/annual.pdf (accessed 1 July 2003).

Jargowsky, Paul A. 2003. *Stunning progress, hidden problems: The dramatic decline of concentrated poverty in the 1990s.* Washington, D.C.: Brookings Institution.

Johnson, Gary R., Richard J. Hauer, and Jill D. Pokorny. 2003. Prevention of hazardous tree defects. In *Urban tree risk management: A community guide to program design and implementation,* ed. Jill D. Pokorny, 117–22. St. Paul: USDA Forest Service.

Jokisch, Brad, and J. Pribilsky. 2002. The panic to leave: Geographic dimensions of recent Ecuadorian emigration. *International Migration* 40: 75–101.

Joubert, Lorraine, and James Lucht. 2000. *Wickford Harbor watershed assessment.* Kingston, R.I.: University of Rhode Island Cooperative Extension. http://www.uri.edu/ce/wq/mtp/wick/repnmaps/FINALWickReport%202.pdf (accessed 10 January 2003).

Jun, M. J. 2004. The effects of Portland's urban growth boundary on urban development patterns and commuting. *Urban Studies* 41: 1333–48.

Kaplan, Rachel. 2001. The nature of the view from home—Psychological benefits. *Environment and Behavior* 33: 507–42.

Kaplan, Robert D. 1996. *The ends of the earth: A journey to the frontiers of anarchy.* New York: Vintage Books.

———. 1997. *The ends of the earth: From Togo to Turkmenistan, from Iran to Cambodia—A journey to the frontiers of anarchy.* New York: Vintage Books.

———. 2000. *The coming anarchy: Shattering the dreams of the post cold war.* New York: Random House.

Kenny, J. T. 1995. Climate, race, and imperial authority: The symbolic landscape of the British hill station in India. *Annals of the Association of American Geographers* 85: 694–714.

———. 1997. Claiming the high ground: Theories of imperial authority and the British hill stations in India. *Political Geography* 16: 655–73.

Kenzer, M. 2000. Healthy cities: A guide to the literature. *Public Health Reports* 115: 279–89.

Keraita, B., P. Drechsel, and P. Amoah. 2003. Influence of urban wastewater on stream water quality and agriculture in and around Kumasi, Ghana. *Environment and Urbanization* 15: 171–78.

Khisty, C. J., and C. K. Ayvalik. 2003. Automobile dominance and the tragedy of the land-use/transport system: Some critical issues. *Systemic Practice and Action Research* 16: 53–73.

Knox, Paul L., and Steven Pinch. 2000. *Urban social geography: An introduction.* Harlow, England: Prentice-Hall.

Kunstler, James Howard. 1993. *The geography of nowhere: The rise and decline of America's man-made landscape.* New York: Simon and Schuster.

Kvorning, Jens. 2002. Copenhagen: Formation, change, and urban life. In *The urban lifeworld: Formation, perception, representation,* ed. Peter Madsen and Richard Plunz, 115–35. London: Routledge.

Laituri, M. 1996. Cross-cultural dynamics in the eco-city: Waitakere City, New Zealand. *Cities* 13: 329–37.

Lalou, Richard, and Thomas K. LeGrand. 1997. Child mortality in the urban and rural Sahel. *Population: An English Selection* 9: 147–68.

Lange, Deborah, and Sue McNeil. 2004. Clean it and they will come? Defining successful brownfield development. *Journal of Urban Planning and Development-Asce* 130: 101–8.

Langeweg, F. 1998. The implementation of agenda 21: 'Our common failure'? *Science of the Total Environment* 218: 227–38.

Law, Neely L., Lawrence E. Band, and J. Morgan Grove. 2004. Nitrogen input from residential lawn care practices in suburban watersheds in Baltimore County, MD. *Journal of Environmental Planning and Management* 47: 737–55.

Lehman, Tim. 1995. *Public values, private lands: Farmland preservation policy, 1933–1985*. Chapel Hill: University of North Carolina Press.

Lemon, James T. 1985. *Toronto since 1918: An illustrated history*. Toronto: Lames Lorimer & Company National Museum of Man, National Museums of Canada.

———. 1996. *Liberal dreams and nature's limits: Great cities of North America since 1600*. Toronto: Oxford University Press.

Lesnikowski, Wojciech G. 1982. *Rationalism and romanticism in architecture*. New York: McGraw-Hill.

Ley, David. 1983. *A social geography of the city*. New York: Harper and Row.

Lindsey, Greg, Maltie Maraj, and SonCheong Kuan. 2001. Access, equity, and urban greenways: An exploratory investigation. *Professional Geographer* 53: 332–46.

Lipset, Seymour Martin. 1959. Some social requisites of democracy: Economic development and political legitimacy. *American Political Science Review* 53: 69–105.

Litfin, Karen T. 1997. Sovereignty in world ecopolitics. *Mershon International Studies Review* 41: 167–204.

Little, Charles E. 1990. *Greenways for America*. Baltimore: Johns Hopkins University Press.

Livernash, Robert, and Eric Rodenburg. 1998. *Population change, resources, and the environment*. New York: Population Reference Bureau.

Lock, Karen, and Henk de Zeeuw. 2001. Mitigating the health risks associated with urban and periurban agriculture. *Urban Agriculture Magazine* 3: 6–8.

Logan, Katharine. 2001. Seaside turns twenty. http://www.architectureweek. com/2001/0919/culture_1–1.html (accessed 29 November 2004).

Los Angeles County Department of Health Services and UCLA Center for Health Policy Research. 2000. The burden of disease in Los Angeles County: A study of the pattern of morbidity and mortality in the county population.

Lund, H. 2003. Testing the claims of new urbanism—Local access, pedestrian travel, and neighboring behaviors. *Journal of the American Planning Association* 69: 414–29.

Lynch, Kevin. 1984. *Good city form*. Cambridge: MIT Press.

MacKellar, F. L. 1997. Population and fairness. *Population and Development Review* 23: 359–76.

Madaleno, I. 2000. Urban agriculture in Belem, Brazil. *Cities* 17: 73–77.

Maisels, Charles Keith. 1990. *The emergence of civilization: From hunting and gathering to agriculture, cities, and the state in the Near East*. London: Routledge.

Malthus, Thomas R. 1798. *An essay on the principle of population and its affects on the future improvement of society, with remarks on the speculations of Mr. Godwin, M. Condorcet and other writers*. London: Printed for J. Johnson.

Marcais, William. 1928. L' Islamisme et la vie urbaine. *L'Academie des Inscriptions et Belles-Lettres, Comptes Rendus* (January–March): 86–100.

Massachusetts Water Resources Authority. 2003. The MWRA sewer system. http://www.mwra.state.ma.us/03sewer/html/sew.htm (accessed 4 December 2003).

McClintock, Jack. 1999. Peter the Great: This guy turns a sleepy azalea park into one of the best botanical gardens in the hemisphere, so now he thinks he can save the world too? *Discover* 20, no. 10. http://www.discover.com/issues/oct-99/departments/featpeter/ (accessed 13 January 2006).

McGee, Terry. 2001. Urbanization takes on new dimensions in Asia's population giants. *Population Today* 29: 1–2.

McGranahan, Gordon, and David Satterthwaite. 2002. Environmental health or ecological sustainability? Reconciling the brown and green agendas in urban development. In *Planning in cities: Sustainability and growth in the developing world*, ed. Roger Zetter and Rodney White, 43–57. London: ITDG.

McInnis, Marvin. 1990. The demographic transition. In *Historical atlas of Canada*. Vol. 3, *Addressing the twentieth century, 1891–1961*, ed. Donald Kerr and Deryck W. Holdsworth, plate 29. Toronto: University of Toronto Press.

McMahon, S. K. 2002. The development of quality of life indicators—A case study from the city of Bristol, U.K. *Ecological Indicators* 21: 177–85.

McShane, Clay. 1988. Urban pathways: The street and highway, 1900–1940. In *Technology and the rise of the networked city in Europe and America*, ed. Joel A. Tarr and Gabriel Dupuy, 67–87. Philadelphia: Temple University Press.

McShane, Clay, and Joel A. Tarr. 1997. The centrality of the horse in the nineteenth-century American city. In *The making of urban America*, ed. Raymond A. Mohl, 105–30. Wilmington, Del.: Scholarly Resources.

Meadows, Robin. 2002. California increases support for agricultural easements. *California Agriculture* 56: 6–7.

Melosi, Martin V. 2000. *The sanitary city: Urban infrastructure in America from colonial times to the present*. Baltimore: Johns Hopkins University Press.

Mercier, Michael E. 2004. The social geography of childhood mortality in Toronto, Ontario, 1901. Ph.D. diss., Department of Geography and Geology, McMaster University.

Mercier, Michael E., and Christopher G. Boone. 2002. Infant mortality in Ottawa, Canada, 1901: Assessing cultural, economic, and environmental factors. *Journal of Historical Geography* 28: 486–507.

Mertz, Lee. 2002. Origins of the interstate. U.S. Department of Transportation, Federal Highway Administration. http://www.fhwa.dot.gov/infrastructure/origin.htm (accessed 13 March 2003).

Mieszkowski, P., and E. S. Mills. 1993. The causes of metropolitan suburbanization. *Journal of Economic Perspectives* 7: 135–47.

Miller, Char. 2003. In the sweat of our brow: Citizenship in American domestic practice during WWII—victory gardens. *Journal of American Culture* 26: 395–410.

Minnesota Safety Council. 2002. Minnesota traffic facts. http://www.mnsafety council.org/facts/f-facts.cfm?FS=19&BAK=%2Fcrosswalk%2Findex.cfm (accessed 28 July 2004).

Mitchell, Don. 2003. *The right to the city: Social justice and the fight for public space.* New York: Guilford.

Modarres, Ali. 2002. Persistent poverty and the failure of area-based initiatives in the U.S. *Local Economy* 17: 289–302.

———. 2005. *Modernizing Yazd: Selective historical memory and the fate of vernacular architecture.* Costa Mesa, Calif.: Mazda Publishers.

Montiel, R. P., and F. Barten. 1999. Urban governance and health development in Leon, Nicaragua. *Environment and Urbanization* 11: 11–26.

Moolgavkar, S. H. 1994. Air pollution and mortality. *New England Journal of Medicine* 330: 1237–38.

Moomaw, R. L., and A. M. Shatter. 1996. Urbanization and economic development: A bias toward large cities? *Journal of Urban Economics* 40: 13–37.

Mothers of East L.A., Santa Isabel. N.d. http://latino.sscnet.ucla.edu/community/intercambios/melasi/ (accessed 8 January 2006).

Mougeot, Luc J.A. 2000. Urban agriculture: Definition, presence, potential, and risks. In *Growing cities, growing food: Urban agriculture on the policy agenda,* ed. N. Bakker, M. Dubbeling, S. Guendel, U. Sabel-Koschella, and H. de Zeeuw, 1–48. DSE.

Muchaal, Pia. 2001. Zoonoses of dairy cattle with reference to Africa. *Urban Agriculture Magazine* 3: 17–19.

Muller, Edward K. 2001. Industrial suburbs and the growth of metropolitan Pittsburgh, 1870–1920. *Journal of Historical Geography* 27: 58–73.

Mumford, Lewis. 1961. *The city in history: Its origins, its transformations, and its prospects.* New York: Harcourt, Brace and World.

Nagdeve, Dewaram, and D. Bharati. 2003. Urban-rural differentials in maternal and child health in Andhra Pradesh, India. *Rural and Remote Health* 3. http://www.regional.org.an/an/rrh/2003/nagdev.htm (accessed 13 January 2006).

Nassau-Suffolk Regional Planning Board. 1978. *The Long Island comprehensive waste treatment management plan.* Hauppauge, N.Y.: Nassau-Suffolk Regional Planning Board.

National Asthma Control Task Force. 2000. The prevention and management of asthma in Canada: A major challenge now and in the future. Ottawa: Health Canada.

National Center for Health Statistics. 2004. Asthma prevalence, health care use, and mortality, 2000–2001. http://www.cdc.gov/nchs/products/pubs/pubd/hestats/asthma/asthma.htm (accessed 2 August 2004).

Natural Resources Defense Council. 2001. Stormwater strategies: Community responses to runoff pollution. http://www.nrdc.org/water/pollution/storm/chap12.asp (accessed 8 January 2006).

———. 2004. Bush mercury policy threatens the health of women and children. http://www.nrdc.org/media/pressreleases/040227.asp (accessed 2 August 2004).

Needell, Jeffrey D. 1984. Making the carioca belle epoque concrete: The urban reforms of Rio de Janeiro under Periera Passos. *Journal of Urban History* 10: 383–422.

Nef, Jorge. 1995. *Human security and mutual vulnerability: An exploration into the global political economy of development and underdevelopment.* Ottawa: International Development Research Centre.

Nevalainen, J., and J. Pekkanen. 1998. The effect of particulate air pollution on life expectancy. *Science of the Total Environment* 217: 137–41.

NewUrbanism.org. n.d. New urbanism. http://www.newurbanism.org/pages/416429/index.htm (accessed 2 July 2003).

Nicholas, David. 1987. *The metamorphosis of a medieval city: Ghent in the age of the Arteveldes, 1302–1390.* Lincoln: University of Nebraska Press.

Nicholas, S. W., B. Jean-Louis, B. Ortiz, M. Northridge, K. Shoemaker, R. Vaughan, M. Rome, G. Canada, and V. Hutchinson. 2005. Addressing the childhood asthma crisis in Harlem: The Harlem Children's Zone Asthma Initiative. *American Journal of Public Health* 95: 245–49.

Nowak, D. J., and Daniel E. Crane. 2002. Carbon storage and sequestration by urban trees in the USA. *Environmental Pollution* 116: 381–89.

Nowak, D. J., M. H. Noble, S. M. Sisinni, and J. F. Dwyer. 2001. People and trees—Assessing the U.S. urban forest resource. *Journal of Forestry* 99: 37–42.

Organisation for Economic Cooperation and Development (OECD). 2002. Road safety performance: Trends and comparative Analysis. OECD Road Transport Research.

Organisation Internationale des Constructeurs d' Automobiles (OICA). 1999. Statistics yearbook: World motor vehicles. Paris: Organisation Internationale des Constructeurs d'Automobiles.

Olmsted Brothers. 1987. *Development of public grounds for Greater Baltimore.* Baltimore: Friends of Maryland's Olmsted Parks and Landscapes/Walsworth Publishing Company. (Orig. pub. 1904.)

Olson, Sherry H. 1979. Baltimore imitates the spider. *Annals of the Association of American Geographers* 69: 557–74.

———. 1997. *Baltimore: The building of an American city.* Baltimore: Johns Hopkins University Press.

Opschoor, H. 2000. The ecological footprint: Measuring rod or metaphor? *Ecological Economics* 32: 363–65.

Orser, W. Edward. 2003. A tale of two park plans: The Olmsted vision for Baltimore and Seattle. *Maryland Historical Magazine* 98: 467–83.

Otterpohl, R., A. Albold, and M. Oldenburg. 1999. Source control in urban sanitation and waste management: Ten systems with reuse of resources. *Water Science and Technology* 39: 153–60.

Owen, David. 2004. Green Manhattan: Everywhere should look more like New York. *New Yorker,* 18 October, 111–23.

Packer, James E. 1967. Housing and population in imperial Ostia and Rome. *Journal of Roman Studies* 59: 80–95.

Palm, Risa. 1990. *Natural hazards: An integrative framework for research and planning.* Baltimore: Johns Hopkins University Press.

Parsons, Kermit C., and David Schuyler, eds. 2002. *From garden city to green city: The legacy of Ebenezer Howard.* Baltimore: Johns Hopkins University Press.

Partlow, Joshua, and Hamil R. Harris. 2004. Twelve empty houses destroyed in Md. blaze: Nearly 20 others damaged in new Charles County development. *Washington Post*, 6 December.

Pasciuti, Daniel, and Christopher Chase-Dunn. 2002. Estimating the population sizes of cities: Urbanization and empire formation project. http://irows.ucr.edu/research/citemp/estcit/estcit.htm (accessed 2 January 2005).

Patterson, John R. 1992. The city of Rome: From republic to empire. *Journal of Roman Studies* 82: 186–215.

Perdue, W. C., L. A. Stone, and L. O. Gostin. 2003. The built environment and its relationship to the public's health: The legal framework. *American Journal of Public Health* 93: 1390–94.

Platt, Rutherford H., Rowan A. Rowntree, and Pamela C. Muick, eds. 1994. *The ecological city: Preserving and restoring urban biodiversity.* Amherst: University of Massachusetts Press.

Pollock, Susan. 1999. *Ancient Mesopotamia: The Eden that never was.* Cambridge: Cambridge University Press.

Population Reference Bureau. 2002. *2002 world population data sheet of the Population Reference Bureau.* http://www.prb.org/pdf/WorldPopulationDS02_Eng.pdf.

Portney, Kent E. 2003. *Taking sustainable cities seriously: Economic development, the environment, and quality of life in American cities.* Cambridge: MIT Press.

Pred, Allan Richard. 1980. *Urban growth and city systems in the United States, 1840–1860.* Cambridge: Harvard University Press.

Pucher, John, and Lewis Dijkstra. 2003. Promoting safe walking and cycling to improve public health: Lessons from the Netherlands and Germany. *American Journal of Public Health* 93: 1509–16.

Pulido, L. 2000. Rethinking environmental racism: White privilege and urban development in Southern California. *Annals of the Association of American Geographers* 90: 12–40.

Pulido, L., S. Sidawi, and R. O. Vos. 1996. An archaeology of environmental racism in Los Angeles. *Urban Geography* 17: 419–39.

Rahardjo, Tjahjono. 2000. The Semarang environmental agenda: A stimulus to targeted capacity building among the stakeholders. *Habitat International* 24: 443–53.

Rails-to-Trails Conservancy. 2004. What we do. http://www.railtrails.org/whatwedo/ (accessed 9 November 2004).

Ramanathan, R. 2001. The long-run behaviour of transport performance in India: A cointegration approach. *Transportation Research Part A* 35: 309–20.

Ratcliffe, Barrie M. 1990. Cities and environmental decline: Elites and the sewage problem in Paris from the mid-eighteenth to the mid-nineteenth century. *Planning Perspectives* 5: 189–222.

Reddy, S. A., and P. Balachandra. 2003. Integrated energy-environment-policy analysis: A case study of India. *Utilities Policy* 11: 59–73.

Redman, Charles, J. Morgan Grove, and Lauren Kuby. 2004. Integrating social science research into the long-term ecological research (LTER) network: Social dimensions of ecological change and ecological dimensions of social change. *Ecosystems* 7: 161–71.

Rees, W. E. 2000. Eco-footprint analysis: Merits and brickbats. *Ecological Economics* 32: 371–74.

Registrar General and Census Commissioner, India. Census of India. 2001. http://www.censusindia.net/ and http://www.censusindia.net/results/provindia3.html (accessed 10 May 2005).

Reid, Donald. 1991. *Paris sewers and sewermen: Realities and representations.* Cambridge: Harvard University Press.

Reisner, Marc. 1993. *Cadillac desert: The American West and its disappearing water.* Vancouver: Douglas and McIntyre.

Relph, Edward. 1987. *The modern urban landscape.* Baltimore: Johns Hopkins University Press.

Rilla, Ellen. 2002. Landowners, while pleased with agricultural easements, suggest improvements. *California Agriculture* 56: 21–25.

Robbins, Paul, A. Polderman, T. Birkenholtz. 2001. Lawns and toxins—An ecology of the city. *Cities* 18, no. 6: 369–80.

Roddick, Jacqueline. 1997. Earth summit north and south: Building a safe house in the winds of change. *Global Environmental Change* 7: 147–65.

Rome, Adam Ward. 2001. *The bulldozer in the countryside: Suburban sprawl and the rise of American environmentalism.* Cambridge: Cambridge University Press.

Rose, Mark H. 1979. *Interstate express highway politics, 1941–1989.* Lawrence: Regents Press of Kansas.

Roseland, M. 1997. Dimensions of the eco-city. *Cities* 14: 197–202.

Rosenzweig, Roy, and Elizabeth Blackmar. 1992. *The park and the people: A history of Central Park.* Ithaca: Cornell University Press.

Rotheroe, N., M. Keenlyside, and L. Coates. 2003. Local agenda 21: Articulating the meaning of sustainable development at the level of the individual enterprise. *Journal of Cleaner Production* 11: 537–48.

Routray, J. K., and A. K. Pradhan. 1989. Slums and development programs in eastern India—A case-study of Cuttack-City. *Habitat International* 13: 99–108.

Ru-Kang, Fang. 1993. The geographical inequalities of mortality in China. *Social Science and Medicine* 36: 1319–23.

Rutsein, Shea O. 2000. Factors associated with trends in infant and child mortality in developing countries during the 1990s. *Bulletin of the World Health Organization* 78: 1256–70.

Rybczynski, Wiltold. 1995. *City life.* New York: Simon and Schuster.

Saudi Arabia Information Resource. N.d. Desalination. http://www.saudinf.com/main/f42.htm (accessed 24 July 2003).

Sauer, Carl Ortwin. 1952. *Agricultural origins and dispersals.* New York: American Geographical Society.

Schellenberger, Michael, and Ted Nordhaus. 2005. The death of environmentalism: Global warming politics in a post-environmental world. *Grist Magazine,* 13 January. http://www.grist.org/news/maindish/2005/01/13/doc-reprint/.

Schmelzkopf, K. 1995. Urban community gardens as contested space. *Geographical Review* 85: 364–81.

Schorske, Carl E. 1981. The Ringstrasse, its critics, and the birth of urban modernism. In Carl E. Schorske, *Fin-de-siècle Vienna: Politics and culture,* 24–115. New York: Vintage Books.

Schueler, Thomas, and Claytor, Richard. 1997. Impervious cover as an urban stream indicator and a watershed management tool. In *Effects of watershed development and management on aquatic ecosystems: Proceedings of an Engineering Foundation workshop*, ed. Larry A. Rosen, 513–29. New York: American Society for Civil Engineers.

Schuyler, David. 2002. Introduction. In *From garden city to green city: The legacy of Ebenezer Howard*, ed. Kermit C. Parsons and David Schuyler, 1–13. Baltimore: Johns Hopkins University Press.

Searns, R. M. 1995. The evolution of greenways as an adaptive urban landscape form. *Landscape and Urban Planning* 33: 65–80.

Sekhar, K. C., N. S. Chary, C. T. Kamala, J. V. Rao, V. Balaram, and Y. Anjaneyulu. 2003. Risk assessment and pathway study of arsenic in industrially contaminated sites of Hyderabad: A case study. *Environment International* 29: 601–11.

1896. The sewers and sewage farms of Berlin. *Engineering News and American Railway Journal* 36: 139–41.

Shrinking Cities. n.d. Shrinking cities: Halle-Leipzig. http://www.shrinkingcities.com/halle_leipzig.0.html (accessed 13 August 2004).

Slicher van Bath, B. H. 1963. *The agrarian history of western Europe, A.D. 500–1850*. London: E. Arnold.

Smart Growth Network. 2004. Smart growth news from around the web. http://www.smartgrowth.org (accessed 7 December 2004).

Smit, Daniel P. 1984. Urban form and automobile externalities: A collective dilemma model. *Land Use Policy* 4: 299–308.

Smith, H., and J. Raemaekers. 1998. Land use pattern and transport in Curitiba. *Land Use Policy* 15, no. 3: 233–51.

Smith, Maf, J. Whitelegg, and Nick Williams. 1998. *Greening the built environment*. London: Earthscan.

Snell, Bradford C. 1974. *American ground transport: A proposal for restructuring the automobile, truck, bus, and rail industries*. Washington, D.C.: U.S. Government Printing Office.

Sokolow, Alvin D. 2002. Agricultural easements limited geographically. *California Agriculture* 56: 15–20.

Sokolow, Alvin D., and Cathy Lemp. 2002. Saving agriculture or saving the environment? *California Agriculture* 56: 9–14.

Spangenberg, Joachim H., and Sylvia Lorek. 2002. Environmentally sustainable household consumption: From aggregate environmental pressures to priority fields of action. *Ecological Economics* 43: 127–40.

Spangenberg, Joachim H., Stefanie Pfahl, and Kerstin Deller. 2002. Towards indicators for institutional sustainability: Lessons from an analysis of agenda 21. *Ecological Indicators* 2: 61–77.

Spirn, Anne Whiston. 1984. *The granite garden: Urban nature and human design*. New York: Basic Books.

Steinberg, F., and L. M. Sara. 2000. The Peru urban management education programme (PEGUP)—Linking capacity building with local realities. *Habitat International* 24: 417–31.

Stone-Lee, Ollie. 2004. Is the green belt getting looser? *BBC News Online*, 12 September.

Szreter, S. 2001. History of demography. In *International encyclopedia of the social and behavioral sciences*, ed. N. J. Smelser and P. B. Baltes, 4:3488–93. Amsterdam: Elsevier.

Takano, T., J. Fu, K. Nakamura, K. Uji, Y. Fukuda, M. Watanabe, and H. Nakajima. 2002. Age-adjusted mortality and its association to variations in urban conditions in Shanghai. *Health Policy* 61: 239–53.

Tarr, Joel A. 1984. The evolution of the urban infrastructure in the nineteenth and twentieth centuries. In *Perspectives on urban infrastructure*, ed. Royce Hanson, 4–65. Washington, D.C.: National Academy Press.

———. 2002. The metabolism of the industrial city—The case of Pittsburgh. *Journal of Urban History* 28: 511–45.

Teaford, Jon C. 1984. *The unheralded triumph: City government in America, 1870–1900*. Baltimore: Johns Hopkins University Press.

Te Lintelo, Dolf, Fiona Marshall, and D. S. Bhupal. 2001. Peri-urban agriculture in Delhi, India. *Food, Nutrition and Agriculture* 29: 4–13.

Tesar, Maria, Thomas G. Reichenauer, and Angela Sessitsch. 2002. Bacterial rhizosphere populations of black poplar and herbal plants to be used for phytoremediation of diesel fuel. *Soil Biology and Biochemistry* 34: 1883–92.

Thomas, David. 1970. *London's green belt*. London: Faber and Faber.

Thornton, Marilza. 2005. Environmental injustice in Brasilia: Who are the people living in Estrurural and why? Master's thesis, Department of Latin American Studies, Ohio University.

Thornton, P. A., and S. Olson. 1991. Family contexts of fertility and infant survival in nineteenth-century Montreal. *Journal of Family History* 16: 401–17.

———. 2001. A deadly discrimination among Montreal infants, 1860–1900. *Continuity and Change* 16: 95–135.

Timaeus, Ian M., and Louisiana Lush. 1995. Intra-urban differentials in child health. *Health Transition Review* 5: 163–90.

Tominaga, K. 2001. East Asia: Environmental issues. In *International encyclopedia of the social and behavioral sciences*, ed. N. J. Smelser and P. B. Baltes, 3943–47. Amsterdam: Elsevier.

Transportation Research Board. 2001. Making transit work: Insight from Western Europe, Canada, and the United States. Washington, D.C.: National Research Council.

Tuason, Julie. 1997. *Rus in Urbe*: The spatial evolution of urban parks in the United States, 1850–1920. *Historical Geography* 25: 124–47.

Twiss, J., J. Dickinson, S. Duma, T. Kleinman, H. Paulsen, and L. Rilveria. 2003. Community gardens: Lessons learned from California healthy cities and communities. *American Journal of Public Health* 93: 1435–38.

UN-HABITAT. n.d. http://www.unhabitat.org/programmes/guo/guo_hsdb4.asp (accessed 25 November 2004).

United Church of Christ and Commission for Racial Justice. 1987. *Toxic wastes and race in the United States: A national report on the racial and socioeconomic characteristics of communities with hazardous waste sites*. New York: United Church of Christ.

United Nations. 2000. World urbanization prospects: The 1999 revision. http://www.un.org/esa/population/publications/wup1999/wup99.htm.

United Nations Development Programme. 1996. *Urban agriculture: Food, jobs and sustainable cities*. Publication Series for Habitat II, vol. 1. New York: United Nations Development Programme.

United Nations Economic and Social Council. 1995. Implication of the recommendations of the international conference on population and development for the work programme on population. Resolution 1995/55. 28 July. http://www.un.org/documents/ecosoc/res/1995/eres1995-55.htm.

United Nations Economic Commission for Europe. 2003. Convention on access to information, public participation in decision-making and access to justice in environmental matters. http://www.unece.org/env/pp/ (accessed 24 January 2004).

United Nations Environment Programme. 2000. International source book on environmentally sound technologies for wastewater and stormwater management. http://www.unep.or.jp/ietc/Publications/TechPublications/TechPub-15/3–6EuropeWest/6–3-2_1.asp (accessed 3 December 2003).

———. 2002. *Global environment outlook 3 (GEO-3)*. London: Earthscan Publications.

United Nations Population Fund. 1996. State of the world population: Changing places; population, development, and the urban future. http://www.unfpa.org/swp/1996/ch2.htm (accessed 13 July 2004).

U.S. Census Bureau. 2004. Mini historical statistics. http://www.census.gov/statab/www/minihs.html (accessed 21 May 2004).

U.S. Department of Agriculture. 2002a. 2002 farm bill NRCS. http://www.nrcs.usda.gov/programs/farmbill/2002/fpprfp.html (accessed 5 May 2004).

———, Office of the Chief Economist. 2002b. USDA agricultural baseline projections to 2011. Washington, D.C.: Interagency Agricultural Projections Committee.

———. 2003. Farmers market nutrition service. http://www.fns.usda.gov/wic/FMNP/default.htm (accessed 28 June 2004).

U.S. Department of Commerce. 2001. Passenger vehicle market: China. http://strategis.ic.gc.ca/epic/internet/inimr-ri.nsf/fr/gr-74200f.html (accessed 10 August 2004).

U.S. Department of Energy. 1995. *Household energy consumption and expenditures, 1993*. Pittsburgh: U.S. Government Printing Office. ftp://ftp.eia.doe.gov/pub/consumption/residential/rx93ce1.pdf (accessed 10 January 2003).

———. 2001. Biodiesel offers fleets a better alternative to petroleum diesel. Technical Assistance Fact Sheet, Office of Energy Efficiency and Renewable Energy. May. http://www.eere.energy.gov/cleancities/blends/pdfs/biodiesel_fs.pdf.

———. Energy efficiency and renewable energy. 2003a. FreedomCAR and fuel initiative. http://www.eere.energy.gov/hydrogenfuel/ (accessed 10 March 2003).

———. 2003b. Weekly retail premium gasoline prices (including taxes). http://www.eia.doe.gov/emeu/international/gas1.html (accessed 24 July 2003).

———. 2004. Mexico: Environmental briefs. http://www.eia.doe.gov/emeu/cabs/mexenv.html (accessed 2 August 2004).

U.S. Department of Transportation. 2002a. Traffic safety facts 2002. Washington, D.C.: National Highway Traffic Safety Administration.

———. 2002b. Traffic safety facts 2002: Children. Washington, D.C.: National Highway Traffic Safety Administration.

———. 2003. 2001 national household survey. http://nhts.ornl.gov/2001/html_files/trends_ver6.shtml (accessed 20 December 2004).

———, Federal Highway Administration. 1999. *1999 status of the nation's high-ways, bridges, and transit: Conditions and performance report*. http://www.fhwa.dot.gov/policy/1999cpr/summary.htm (accessed 2 January 2003).

U.S. Environmental Protection Agency. 1990. Overview, the Clean Air Act amendments of 1990. http://www.epa.gov/oar/caa/overview.txt (accessed 2 August 2004).

———. 2000. Sulfur dioxide: How sulfur dioxide affects the way we live and breathe. Washington, D.C.: Office of Air Planning and Standards.

———. 2004a. Air trends: Six principal pollutants. http://www.epa.gov/airtrends/sixpoll.html (accessed 2 August 2004).

———. 2004b. The ozone report: Measuring progress through 2003. Washington, D.C.: Environmental Protection Agency.

U.S. General Accounting Office. 1983. Siting of hazardous waste landfills and their correlation with racial and economics status of surrounding communities. Washington, D.C.: U.S. General Accounting Office.

Valentin, Anke, and Joachim H. Spangenberg. 2000. A guide to community sustainability indicators. *Environmental Impact Assessment Review* 20: 381–92.

Van Kooten, G. C., and E. H. Bulte. 2000. The ecological footprint: Useful science or politics? *Ecological Economics* 32: 385–89.

Vance, James E. 1990. *The continuing city: Urban morphology in Western civilization*. Baltimore: Johns Hopkins University Press.

Visaria, Leela, and Parvin Visaria. 1995. India's population in transition. *Population Bulletin* 50: 1–51.

Vitousek, Peter, John Aber, Robert W. Howarth, Gene E. Likens, Pamela A. Matson, David W. Schindler, William H. Schlesinger, G. David Tilman. 1997. Human alteration of the global nitrogen cycle: Causes and consequences. *Ecological Applications* 7: 737–50.

Von Grunebaum, Gustave. 1955. The structure of the Muslim town. In *Islam: Essays in the nature and growth of cultural tradition*. Memoir no. 81. Ann Arbor: American Anthropological Association.

Vostal, J. J. 1999. Fine particles in the ambient air and urban mortality: The epidemiological evidence and pathogenetic mechanisms. *Journal of Aerosol Science* 30: S795–96.

Wachs, Martin. 1996. The evolution of transportation policy in Los Angeles: Images of past policies and future prospects. In *The city: Los Angeles and urban theory at the end of the twentieth century*, ed. Allen John Scott and Edward W. Soja, 106–59. Berkeley and Los Angeles: University of California Press.

Wackernagel, Mathis, and William E. Rees. 1996. *Our ecological footprint: Reducing human impact on the earth*. Gabriola Island, British Columbia: New Society Publishers.

Ward, David. 1971. *Cities and immigrants: A geography of change in nineteenth-century America*. New York: Oxford University Press.

————. 1989. *Poverty, ethnicity, and the American city, 1840–1925: Changing conceptions of the slum and the ghetto*. Cambridge: Cambridge University Press.

Ward, Lorne H. 1990. Roman population, territory, tribe, city, and army size from the republic's founding to the Veientane War, 509 B.C.–400 B.C. *American Journal of Philology* 111: 5–39.

Ward, Stephen. 2002. Ebenezer Howard: His life and times. In *From garden city to green city: the legacy of Ebenezer Howard*, ed. Kermit C. Parsons and David Schuyler, 14–37. Baltimore: Johns Hopkins University Press.

Waring, George E., Jr. 1883. Sanitary crainage. *North American Review* 137: 57–67.

Warner, Sam Bass, Jr. 1978. *Streetcar suburbs: The process of growth in Boston (1870–1900)*. Cambridge: Harvard University Press.

Werna, E., T. Harpham, I. Blue, and G. Goldstein. 1999. From healthy city projects to healthy cities. *Environment and Urbanization* 11: 27–39.

Wheeler, Stephen M. 2002. Planning for metropolitan sustainability. *Journal of Planning Education and Research* 20: 135–45.

Wheeler, Stephen M., and Timothy Beatley. 2004. *The sustainable urban development reader*. London: Routledge.

Whitby, K. T., W. E. Clark, V. A. Marple, G. M. Sverdrup, G. J. Sem, K. Willeke, B.Y.H. Liu, and D.Y.H. Pui. 1975. Characterization of California aerosols—I. Size distributions of freeway aerosol. *Atmospheric Environment* 95: 463–82.

White, Gilbert F. 1958. *Changes in urban occupance of flood plains in the United States*. Chicago: University of Chicago Press.

White, Morton Gabriel, and Lucia White. 1962. *The intellectual versus the city, from Thomas Jefferson to Frank Lloyd Wright*. Cambridge: Harvard University Press.

Whitehand, J.W.R. 1977. The basis for an historico-geographical theory of urban form. *Transactions of the Institute of British Geographers* 2, no. 3: 400–416.

Williams, Michael. 1990. *Americans and their forests: A historical geography*. Cambridge: Cambridge University Press.

Wilson, R., and J. Spengler, eds. 1996. *Particles in our air: Concentrations and health effects*. Cambridge: Harvard University Press.

Wilson, William H. 1989. *The city beautiful movement*. Baltimore: Johns Hopkins University Press.

Wiseman, T. P. 1969. The census in the first century B.C. *Journal of Roman Studies* 59: 59–75.

Wittfogel, Karl A. 1956. The hydraulic civilizations. In *Man's role in changing the face of the earth*, ed. William L. Thomas Jr., 152–64. Chicago: University of Chicago Press.

————. 1957. *Oriental despotism: A comparative study of total power*. New Haven: Yale University Press.

Wolf, Kathy L. Urban nature benefits: Psycho-social dimensions of people and plants. Human Dimensions of the Urban Forest Fact Sheet #1. Seattle: University of Washington, Center for Urban Horticulture. http://www.cfr.washington.edu/research.envmind/HumanBens/PsychBens-FS1.pdf (accessed 3 January 2002).

Woods, Robert. 2000. *The demography of Victorian England and Wales*. Cambridge: Cambridge University Press.

World Bank. 2000. *World development report, 2000/2001: Attacking poverty.* New York: Oxford University Press.

———. 2004. World development indicators. http://devdata.worldbank.org/dataonline/ (accessed 30 June 2004).

World Commission on Environment and Development. 1987. *Our common future.* New York: World Commission on Environment and Development.

World Health Organization. 1995. Global initiative for asthma: Global strategy for asthma management and prevention. NHLBI/WHO Workshop Report. Washington, D.C.: National Institutes of Health.

———. 2003a. Influenza fact sheet number 211. http://www.who.int/mediacentre/factsheets/fs211/en/ (accessed 30 June 2004).

———. 2003b. Maternal and child health information and analysis: Niger. http://www.who.int/trade/en/PRSP43.pdf (accessed 14 July 2004).

———. 2004a. Healthy cities and networks. http://www.who.dk/healthycities/ CitiesAndNetworks/20010828_1 (accessed 20 July 2004).

———. 2004b. Status report on macroeconomics and health: Nepal. http://www.who.int/macrohealth/action/en/rep04_nepal.pdf.

Wright, H. E. 1978. Toward an explanation of the origin of the state. In *Origins of the state: The anthropology of political evolution,* ed. Ronald Cohen and Elman R. Service, 49–68. Philadelphia: Institute for the Study of Human Issues.

1956. Ye olde English green belt. *Journal of the Town Planning Institute* 42: 68–69.

Index

percolation: impervious surfaces and, 101, 125; reinforced turf and, 104–5; sewage treatment and, 97; wetlands and, 125
Persian: attack of Miletus, 9; early cities, 6, 22; Empire, 7
pesticides, 86, 88, 91
Philadelphia: transport of goods to, 79; urban planning in, 27
Philippines, population growth in, 45
Phnom Penh: transportation in, 100
Phoenix: artificial wetlands in, 132; traffic fatalities in, 149; urban ecology of, xii
phytoremediation, 126
Pittsburgh: compared to Detroit, 131; shipping from, 79
Plato: ideas on the city, 9; secular-rational philosophy and, 26
plaza: in early Roman cities, 19; in front of Saint Peter's dome, 28, in Middle Eastern cities, 24–25; shopping, 183
policy, environmental: brown agenda and, xiii; combined with population policy, 50, 72–73, 76; ecological footprint and, 82; green agenda and, xiii; greenbelts and, 178; inadequate, 58; and sustainability, 188–89; United Nations' Agenda and, 62, 73–74; United Nations Environment Programme and, 51
pollution, nonpoint: 103, 157
population: of ancient cities, 39, 40–41; complexity of, 38; and the demographic transition model, 43–45; density and infrastructure 127–31, 151; and the green revolution, 87–88; and globalization, 76; and health, 52–60, 136, 138, 144, 151; historical demography and, 4–45; historical estimates of, 43; international policies on, 70–76; and relationship to income, 45–52; and resource consumption, 40, 45–52, 48, 49, 81, 83, 86; and sustainability 49, 61–63, 65, 71–73; and urbanization rates, 38, 79; and urban sprawl, 83, 120, 161, 169; and women's rights, 72–73
port cities: Baltimore as and health implications, 137; in Europe, 28
Portland: as livable city, 188; streetcars in, 118; urban growth boundary in, 177–79; walkability of, 152
Porto Alegre: infant mortality rates in shantytowns of, 137
postmodern: urban form, 35–37
poverty: and health, 33, 54, 128, 137; and immigrants, 32; in India, 57; in industrial

cities, 31–32; and infrastructure, 133, 140; and Malthus, 41; in modern cities, 53, 58, 61; in medieval cities, 18; pollution of, 82; and population issues, 73, 75; and sustainable development, xiii
prevailing wind: street construction in consideration of, 13
Priene, 10
progressive era, 32; playgrounds and, 164–66
public baths: in Ancient Rome, 15; in Middle Eastern cities, 24
public space: new urbanism and, 182; in Rome, 16; in Ur, 6; in Vienna, 34
public works, 95–96
public works: in ancient cities, 7, 17, 18; as compared to infrastructure, 96; in modern cities, 95; privatization of, 96; sanitation and, 128

Qatar: gross national income and urbanization in, 46

rationalism: industrial cities and scientific, 30; park planning and, 163; urban morphology and, 29–30
Regensburg, 19
reinforced turf, 104
resilience, 2
retailing: mixed land use and, 152, 186; street trees, 105; and suburbanization, 117, 119; in suburbs, 77, 154, 170, 183; and transportation improvements and, 106
Rio de Janeiro: parks in, 164: UN summit in, 72–74
Rishi Valley: as sustainability example, 65
Rome, Ancient: health problems of, 136; land use of, 15–17, planning and growth of, 14–18, 28; sanitary sewers in, 136
Russia, demographic transition of, 44

Samarkand, 23
San Francisco: conurbation of, 83; farmland preservation near, 84–85; health conditions of, 136; streetcars in, 118; traffic fatalities in, 149; walkability of, 152
Santa Barbara: desalination plant in, 123
São Paulo: as megacity, 61
Saudi Arabia: desalination plants in, 123
Sauer, Carl: and agricultural origins, 2
Senegal: mortality rates in, 57
Seoul: greenbelt around, 178

CHRISTOPHER G. BOONE is Associate Professor and holds joint appointments in the International Institute for Sustainability as well as the School of Human Evolution and Social Change, Arizona State University.

ALI MODARRES is the Associate Director of the Edmund G. "Pat" Brown Institute of Public Affairs and a Professor in the Department of Geography and Urban Analysis at California State University, Los Angeles.